Introduction to Smart Antennas

Introduction to Smart Antennas

Constantine A. Balanis, Panayiotis I. Ioannides

ISBN: 978-3-031-00405-6	paperback
ISBN: 978-3-031-00405-6	paperback
ISBN: 978-3-031-01533-5	ebook
ISBN: 978-3-031-01533-5	ebook

DOI: 10.1007/978-3-031-01533-5

A Publication in the Springer series
SYNTHESIS LECTURES ON ANTENNAS #5

Lecture #5
Series Editor: Constantine A. Balanis, Arizona State University

First Edition
10 9 8 7 6 5 4 3 2 1

Introduction to Smart Antennas

Constantine A. Balanis
Panayiotis I. Ioannides
Department of Electrical Engineering
Arizona State University

SYNTHESIS LECTURES ON ANTENNAS #5

ABSTRACT

As the growing demand for mobile communications is constantly increasing, the need for better coverage, improved capacity and higher transmission quality rises. Thus, a more efficient use of the radio spectrum is required. *Smart antenna systems* are capable of efficiently utilizing the radio spectrum and, thus, is a promise for an effective solution to the present wireless systems' problems while achieving reliable and robust high-speed high-data-rate transmission. The purpose of this book is to provide the reader a broad view of the system aspects of smart antennas. In fact, smart antenna systems comprise several critical areas such as individual antenna array design, signal processing algorithms, space-time processing, wireless channel modeling and coding, and network performance. In this book we include an overview of smart antenna concepts, introduce some of the areas that impact smart antennas, and examine the influence of interaction and integration of these areas to *Mobile Ad-Hoc Networks*. In addition, the general principles and major benefits of using space–time processing are introduced, especially employing multiple-input multiple-output (MIMO) techniques.

KEYWORDS

Adaptive arrays, Switched-beam antennas, Phased array, SDMA, Mutual coupling, Direction of arrival, Adaptive beamforming, Channel coding, MANET, Network throughput, Space–time processing.

Contents

CHAPTER 1

Introduction

In recent years a substantial increase in the development of broadband wireless access technologies for evolving wireless Internet services and improved cellular systems has been observed [1]. Because of them, it is widely foreseen that in the future an enormous rise in traffic will be experienced for mobile and personal communications systems [2]. This is due to both an increased number of users and introduction of new high bit rate data services. This trend is observed for second-generation systems, and it will most certainly continue for third-generation systems. The rise in traffic will put a demand on both manufacturers and operators to provide sufficient *capacity* in the networks [3]. This becomes a major challenging problem for the service providers to solve, since there exist certain negative factors in the radiation environment contributing to the limit in capacity [4].

A major limitation in capacity is co-channel interference caused by the increasing number of users. The other impairments contributing to the reduction of system performance and capacity are multipath fading and delay spread caused by signals being reflected from structures (e.g., buildings and mountains) and users traveling on vehicles. To aggravate further the capacity problem, in 1990s the Internet gave the people the tool to get data on-demand (e.g., stock quotes, news, weather reports, e-mails, etc.) and share information in real-time. This resulted in an increase in airtime usage and in the number of subscribers, thus saturating the systems' capacity.

Wireless carriers have begun to explore new ways to maximize the spectral efficiency of their networks and improve their return on investment [5]. Research efforts investigating methods of improving wireless systems performance are currently being conducted worldwide. The deployment of *smart antennas* (SAs) for wireless communications has emerged as one of the leading technologies for achieving high efficiency networks that maximize capacity and improve quality and coverage [6]. Smart Antenna systems have received much attention in the last few years [6–11] because they can increase system capacity (very important in urban and densely populated areas) by dynamically tuning out interference while focusing on the intended user [12, 13] along with impressive advances in the field of digital signal processing.

Selected control algorithms, with predefined criteria, provide adaptive arrays the unique ability to alter the radiation pattern characteristics (nulls, sidelobe level, main beam direction, and beamwidth). These control algorithms originate from several disciplines and target specific applications (e.g., in the field of seismic, underwater, aerospace, and more recently cellular communications) [14]. The commercial introduction of SAs is a great promise for big increase in system performance in terms of capacity, coverage, and signal quality, all of which will ultimately lead to increased spectral efficiency [14].

As the necessity of exchanging and sharing data increases, users demand ubiquitous, easy connectivity, and fast networks whether they are at work, at home, or on the move. Moreover, these users are interested in interconnecting all their personal electronic devices (PEDs) in an *ad hoc* fashion. This type of network is referred to as Mobile Ad hoc NETwork (MANET), and it is beginning to emerge using Bluetooth™ technology. Bluetooth™ is a short-range, low-power radio link (10–100 m) that allows two or more Bluetooth™ devices to form a communication channel and exchange data [15, 16]. Because Bluetooth™ uses an omnidirectional antenna (operating in the unlicensed 2.4 GHz industrial, scientific, and medical (ISM) band), it lacks the ability to steer the radiation beam toward the intended users and form nulls to cancel jammers. This limits the overall system capacity or network *throughput* of MANETs. Furthermore, because of the omnidirectional antenna, battery life in PEDs is reduced since energy is radiated everywhere and not just toward the desired user. Consequently, the benefits provided by smart antennas would enhance the overall performance of MANETs [17].

Current trends concentrate on space–time processing and coding, a technique that promises to greatly improve the performance in wireless networks by using multiple antennas at the transmitter and the receiver [18]. Space–time processing can be viewed as an evolution of the traditional array signal processing techniques such as antenna array and beamforming. Operating simultaneously on multiple sensors, space–time receivers process signal samples both in time and space, thereby improving resolution, interference suppression, and service quality. Sophisticated space–time processing methods applied to multiple-input multiple-output (MIMO) systems are expected to provide great capacity and data rate increases in cellular systems and wireless local area networks.

This book is organized as follows: in Chapter 2 an overview of wireless communication systems is presented, a requisite to analyze smart antenna systems. Following this, a chapter on antenna arrays and diversity techniques is included that describes antenna properties and classifies them according to their radiation characteristics. In Chapter 4, the functional principles of smart antennas are analyzed, different smart antenna configurations are exhibited and the benefits and drawbacks concerning their commercial introduction are highlighted. Chapter 5 deals with different methods of estimating the direction of arrival. The more accurate this estimate is, the better the performance of a smart antenna system. Chapter 6 is devoted to beamforming techniques through which the desired radiation patterns of the adaptive arrays are achieved. The succeeding chapter presents the results of a project that examines and integrates antenna design, adaptive algorithms and network *throughput*. Chapter 8 is devoted to space–time processing techniques. The fundamental principles are analyzed and, through experimental results, the enormous improvements in data rates and capacities realized with MIMO systems are demonstrated. Before the book is concluded, commercial efforts and products of smart antenna are briefly reviewed in Chapter 9.

This book is a comprehensive effort on smart antenna systems and contains material extracted from various sources. The authors have attempted to indicate, in the respective chapters of the book, the sources from which the material was primarily derived and its development based upon. In particular, the authors would like to acknowledge that major contributions were derived from many references, especially [17, 19–29]. Also, the authors have contacted most of the primary authors of these references, who gracefully and promptly responded favorably. In fact, some of the authors provided expeditiously figures and data included in this book. Acknowledgement of the sources is indicated in the respective figures.

CHAPTER 2

Mobile Communications Overview

In this chapter, a brief overview of mobile communications is presented to understand its functional principles and introduce the necessary terminology for the rest of this book.

2.1 GENERAL DESCRIPTION

All communication systems have fundamentally the same goal: to pass along the maximum amount of information with the minimum number of errors [19]. Modern digital wireless communications systems are no exception. These systems can usually be separated into several elements as indicated by Fig. 2.1. Given any digital input, the source encoder eliminates redundancy in the information bits, thus maximizing the amount of the useful information transferred in the communications system [19]. The output of the source generator is processed by the channel encoder, which incorporates error control information in the data to minimize the probability of error in transmission.

The output of the channel encoder is further processed by the Digital Signal Processing unit, in order to allow simultaneous communication of many users. An example of this would be digital beamforming, which by using the geometric properties of the antenna array, is able to concentrate signals from multiple users in different desired directions, allowing more users to be served by the system. The generated data stream is then processed by the modulator which is responsible to shift the baseband signal at its input into the band-pass version at the output, due to the bandwidth constraints of the communication system [19]. The information sequence generated at the output of the modulator is then fed into the antenna array and transmitted through the wireless channel.

On the other end of the radio channel, the reverse procedure takes place. The demodulator down converts the signals from different users collected by the receiver antenna into their baseband equivalent. The Digital Signal Processor then separates the different signals that come from different users. The channel decoder detects and corrects, if possible, errors that are caused due to propagation through the physical channel. Following that, the source decoder restores the actual data sequence from its compressed version. The entire procedure aims to recover the information transmitted on the other end of the physical channel, with the least possible number of errors.

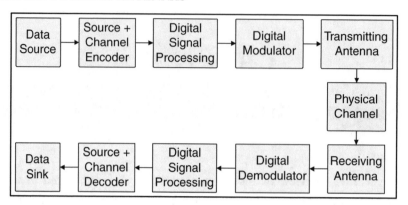

FIGURE 2.1: Elements of a communications system [19].

2.2   CELLULAR COMMUNICATIONS OVERVIEW

The wireless communications era began around 1895 when Guglielmo Marconi demonstrated the use of radio waves to communicate over large distances. Cellular is currently one of the fastest growing and most demanding telecommunications applications. Today, it represents the dominant percentage of all new telephone subscriptions around the world. During the early part of this decade, the number of mobile cellular subscribers has surpassed that of conventional fixed lines [30]. In many parts of the world, cell phone penetration is already over 100% and the market is still growing. According to the latest figures from Wireless Intelligence (WI) [31], the venture between Ovum and the GSM Association that focuses on market data and analysis on the global wireless industry, worldwide growth is currently running at over 40 million new connections per month—the highest volume of growth the market has ever seen. Overall, world market penetration is expected to rise from an estimated 41% at the end of 2006 to 47% by the end of 2007, on a track to hit the landmark of 3 billion cellular connections! However, as Wireless Intelligence says, the number of cellular connections does not represent the number of cellular users, since many subscribers have more than one cellular connection and, in addition, these figures include accounts that may no longer be active. In general, subscriber growth is especially strong in Asia, where penetration rates are still low, followed by the Americas while the saturated Western European market is stagnant [32]. The charts in Fig. 2.2 graph Micrologic Research's [33] estimates (a) of the annual worldwide cellular telephone sales and (b) worldwide number of cellular subscribers from 1998 to 2006.

2.3   THE EVOLUTION OF MOBILE TELEPHONE SYSTEMS

The concept of cellular service is the use of low-power transmitters where frequencies can be reused within a geographic area. However, the Nordic countries were the first to introduce

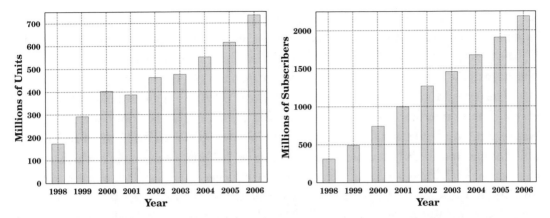

FIGURE 2.2: (a) Annual worldwide cellular handset shipments and (b) worldwide number of cellular subscribers [34].

cellular services for commercial use with the introduction in 1981 of the Nordic Mobile Telephone (NMT).

Cellular systems began in the United States with the release of the advanced mobile phone service (AMPS) system in 1981. The AMPS standard was adopted by Asia, Latin America, and Oceanic countries, creating the largest potential market in the world for cellular technology [35].

In the early 1980s, most mobile telephone systems were analog rather than digital, like today's newer systems. One challenge facing analog systems was the inability to handle the growing capacity needs in a cost-efficient manner. As a result, digital technology was welcomed. The advantages of digital systems over analog systems include ease of signaling, lower levels of interference, integration of transmission and switching, and increased ability to meet capacity demands [35].

GSM, which was first introduced in 1991, is one of the leading digital cellular systems. Today, it is the *de facto* wireless telephone standard in Europe, and it is widely used in Europe and other parts of the world.

CDMA system was first standardized in 1993. CDMA refers to the original ITU IS-95 (CDMA) wireless interface protocol and is considered a second-generation (2G) mobile wireless technology which was commercially introduced in 1995. It quickly became one of the world's fastest-growing wireless technologies.

In 1999, the International Telecommunications Union selected CDMA as the industry standard for new "third-generation" (3G) wireless systems. Many leading wireless carriers are now building or upgrading to 3G CDMA networks in order to provide more capacity for voice traffic, along with high-speed data capabilities [36]. The new version of CDMA, also known as CDMA2000 or IS-2000, is both an air interface and a core network solution for delivering the services that customers are demanding today [37]. A key component of CDMA2000 is its ability to support the full demands of advanced 3G services such as multimedia and other IP-based services. CDMA2000 is the ideal solution for wireless operators who want to take advantage of the new market dynamics created by mobility and the Internet [37].

Universal Mobile Telecommunications System (UMTS) is an evolution of the GSM system. The air interface has been changed from a Time Division Multiple Access (TDMA) based system to a Wideband Code Division Multiple Access (W-CDMA) based air interface. This change was needed to achieve the data rate of 2 Mbps to the mobile which is a 3G requirement [38]. Besides voice and data, UMTS will deliver audio and video to wireless devices anywhere in the world through fixed, wireless, and satellite systems. The UMTS system will serve most of the European countries. Table 2.1 charts the worldwide development of Mobile Telephone Systems.

2.4   THE FRAMEWORK

Wireless communication systems usually perform duplex communication between two points [1]. These two points are usually defined as the Base Station (BS) and the Mobile

TABLE 2.1: The Development of Mobile Telephone Systems [35]

YEAR	MOBILE SYSTEM
1981	Nordic Mobile Telephone (NMT) 450
1983	American Mobile Phone System (AMPS)
1985	Total AccessCommunication System (TACS)
1986	Nordic Mobile Telephony (NMT) 900
1991	American Digital Cellular (ADC)
1991	Global System for Mobile Communication (GSM)
1992	Digital Cellular System (DCS) 1800
1993	CDMA One
1994	Personal Digital Cellular (PDC)
1995	PCS 1900-Canada
1996	PCSóUnited States
2000	CDMA2000
2005	UMTS

Station (MS). The data communication from the BS to the MS is usually referred to as the downlink or forward channel. Similarly, the data communication from the MS to the BS is usually referred to as the uplink or reverse channel. Two systems can exist in the downlink: an antenna system for transmission at the BS and another antenna system for reception at the MS. Additionally, there can be two systems in the uplink: transmission at the MS and reception at the BS [1]. An example of such a system is illustrated in Fig. 2.3.

The cellular telephone system provides a wireless connection to the Public Switched Telephone Network (PSTN) for any user in the radio range of the system [39]. It consists of

- Mobile stations
- Base stations, and
- Mobile Switching Center (MSC).

The base station is the bridge between the mobile users and the MSC via telephone lines or microwave links [39]. The MSC connects the entire cellular system to the PSTN in the cellular system. Fig. 2.4 provides a simplified illustration how a cellular telephone system works.

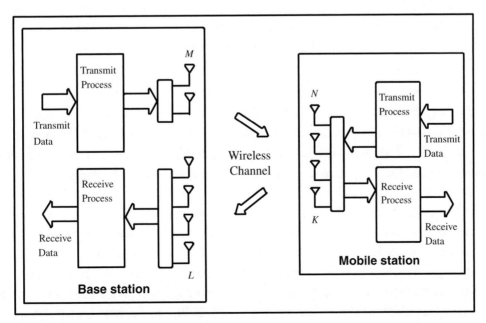

FIGURE 2.3: A general antenna system for broadband wireless communications [1].

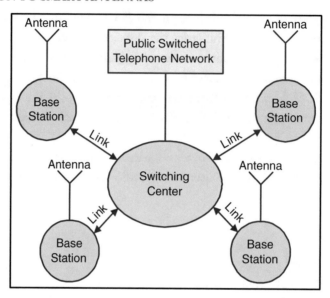

FIGURE 2.4: A typical setup of a base mobile system [40].

2.5 CELLULAR RADIO SYSTEMS: CONCEPTS AND EVOLUTION

Maintaining capacity has always been a challenge as the number of services and subscribers increased. To achieve the capacity demand required by the growing number of subscribers, cellular radio systems had to evolve throughout the years. To justify the need for smart antenna systems in the current cellular system structure, a brief history in the evolution of the cellular radio systems is presented. For in-depth details, the reader is referred to [13, 40, 41].

2.5.1 Omnidirectional Systems and Channel Reuse

Since the early days, system designers knew that capacity was going to be a problem, especially when the number of channels or frequencies allocated by the Federal Communications Commission (FCC) was limited. Therefore, to accommodate the huge number of subscribers and achieve the required capacity, a suitable cellular structure had to be designed. The dominant concept is that the capacity may only be increased by using each traffic channel to carry many calls simultaneously [40]. One way to accomplish this is to use the same channel over and over. To do so, mobile phones using the same radio channel have to be placed sufficiently apart from each other in order to avoid disturbance. Cellurization consists of breaking up a large geographical service area into smaller areas, referred to as *cells*, each of which can use a portion of the available bandwidth (*frequency reuse*), thus making it possible to provide wireless links to many users despite the limited spectrum [42]. Cells, usually, have irregular shapes and dimensions. The shape is determined largely by the terrain and man-made features. Depending

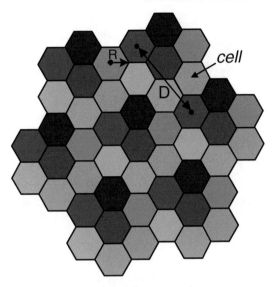

FIGURE 2.5: Typical cellular structure with 7 cells reuse pattern.

on their size, cells can be classified as macrocells (where the base station has sufficient transmit power to cover areas of radius 1–20 km), microcells (areas of 0.1 to 1 km in radius), and picocells (indoor environment) [42]. A minimum distance between two cells using identical channels is required, known as the *channel reuse distance*. This is also known as *channel reuse via spatial separation* [43]. The capacity of the system depends on this distance. An example of such a structure is depicted in Fig. 2.5.

In Fig. 2.5, each hexagonal area with different shade represents a small geographical area named *cell* with maximum radius R [44]. At the center of each cell resides a base station equipped with an omnidirectional antenna with a given band of frequencies. Base stations in adjacent cells are assigned frequency bands that contain completely different frequencies than neighboring cells. By limiting the coverage area within the boundaries of a cell, the same band of frequencies may be used to cover different cells that are separated from each other by distances large enough (indicated as D in Fig. 2.5) to keep interference levels below the threshold of the others. The design process of selecting and allocating the same bands of frequencies to different cells of cellular base stations within a system is referred to as *frequency reuse* or *channel reuse* [41]. This is shown in Fig. 2.5 by repeating the shaded pattern or clusters [13]; cells having the same shaded pattern use the same frequency bandwidth. In the first cellular radio systems deployed, each base station was equipped with an omnidirectional antenna [4]. Because only a small percentage of the total energy reached the desired user, the remaining energy was wasted and polluted the environment with interference. As the number of users increased, so did the interference, thereby reducing capacity. An immediate solution to this

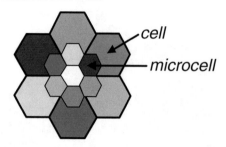

FIGURE 2.6: Cell-splitting.

problem was to subdivide a cell into smaller cells; this technique is referred to as *cell splitting* [44].

2.5.2   Cell Splitting

Cell-splitting [44], as shown in Fig. 2.6, subdivides a congested cell into smaller cells called *microcells*, each with its own base station and a corresponding reduction in antenna height and transmitter power. Cell-splitting improves capacity by decreasing the cell radius R and keeping the D/R ratio unchanged; D is the distance between the centers of the clusters. The disadvantages of cell-splitting are costs incurred from the installation of new base stations, the increase in the number of *handoffs* (the process of transferring communication from one base station to another base station when the mobile unit travels from one cell to another), and a higher processing load per subscriber.

2.5.3   Sectorized Systems

As the demand for wireless service grew even higher, the number of frequencies assigned to a cell eventually became insufficient to support the required number of subscribers. Thus, a cellular design technique was needed to provide more frequencies per coverage area. Sectorized systems subdivide the traditional cellular area into sectors that are covered using directional antennas at the same base station, as shown in Fig. 2.7.

This technique is referred to as *cell-sectoring* [41] where a single omnidirectional antenna is replaced at the base station with several directional antennas. Operationally, each sector is treated as a different cell in the system, the range of which, in most cases, can be greater than in the omnidirectional case (roughly 35% greater), since the transmission power is focused to a smaller area [20].

Sectorized cells can increase the efficient use of the available spectrum by reducing the interference presented by the base station and its users to the rest of the network, and they are widely used for this purpose. Most systems in commercial service today employ three sectors, each one with 120° coverage. Although larger numbers of sectors are possible, the number of

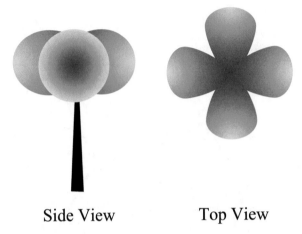

Side View   Top View

FIGURE 2.7: Sectorized antenna system and coverage pattern [20].

antennas and base station equipment become prohibitively expensive for most cell sites [45]. Fig. 2.8 shows a system that employs the 120° type of cell sectorization.

In sectoring, capacity is improved while keeping the cell radius unchanged and reducing the D/R ratio. In other words, capacity improvement is achieved by reducing the number of cells and, thus, increasing the frequency reuse. However, in order to accomplish this, it is necessary to reduce the relative interference without decreasing the transmitting power. The co-channel interference in such cellular system is reduced since only two neighboring cells interfere instead

FIGURE 2.8: Sectorized cellular network employing three sectors, each one covering 120° field of view.

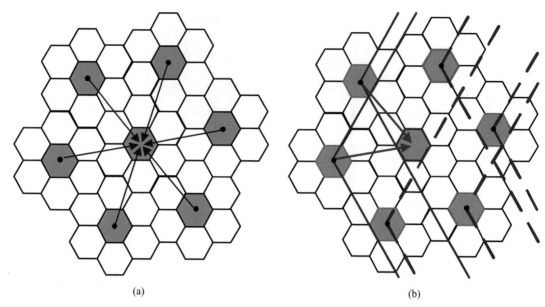

(a)   (b)

FIGURE 2.9: Co-channel interference comparison between (a) omnidirectional and (b) sectorized systems.

of six for the omnidirectional case [44, 46] as shown in Fig. 2.9. Increasing the number of sectors in a CDMA system has been a technique useful of increasing the capacity of cell sites [47]. Theoretically, the increase in capacity is proportional to the number of sectors per cell [48]. The penalty for improved signal-to-interference (S/I) ratio and capacity is an increase in the number of antennas at the base station, and a decrease in *trunking efficiency* [13, 46] due to channel sectoring at the base station. Trunking efficiency is a measure of the number of users that can be offered service with a particular configuration of fixed number of frequencies.

2.6   POWER CONTROL

Power control is a technique whereby the transmit power of a base station or handset is decreased close to the lowest allowable level that permits communication [45]. Due to the logarithmic relationship between the capacity of the wireless link and the signal-to-interference-and-noise ratio (SINR) at the receiver [49], any attempt to increase the data rate by simply transmitting more power is extremely costly. Furthermore, increases in power scales up both the desired signals and their mutual interference [28]. Therefore, once a system has become limited by its own interference, power increase is useless. Since mature systems are designed in a way to achieve maximum capacity, it is the power itself, in the form of interference, that ultimately limits their performance [50]. As a result, power must be carefully controlled and allocated to enable the coexistence of multiple geographically dispersed users operating under various conditions [28] and has been a topic of active research. For example, both GSM and CDMA systems use power control on both uplink and downlink. Particularly, CDMA systems require fast and precise power control since many users share the same RF spectrum, and the system capacity is thus highly sensitive to inadequate interference control [45].

2.6.1   Spectral Efficiency

Another effective way to improve the data rate is to increase the signal bandwidth along with power increase. However, the radio spectrum is not an abundant resource in the frequencies of interest. Moreover, increasing the signal bandwidth beyond the coherence bandwidth results in frequency selectivity and degradation in the transmission quality. *Spectral efficiency*, defined as the ratio of capacity per unit bandwidth, measures the ability of a wireless system to deliver information with a given amount of radio spectrum and provides another key metric of the wireless system's quality. It determines the amount of radio spectrum required to provide a given service (e.g., 10 Kbps voice service or 100 Kbps data service) and the number of base stations required to deliver that service to the end-users. In the latter years of deployment, when subscriber penetration is high, it is, consequently, one of the primary determinants of system economics. Spectral efficiency is measured in units of bits/second per Hertz/cell (b/s/Hz/cell). It determines the total *throughput* each base station (cell or sector) can support with a given amount of spectrum. The appearance of a "per cell" dimension in measuring spectral efficiency may seem surprising, but the *throughput* of a particular base station of a cellular network is almost always substantially less than that of a single cell in isolation. This difference is attributed to the self-interference generated in the network.

In a cellular system, the radio communication between a user and a base station generates radio energy that is detectable in places other than the immediate vicinity of the user, the base station and an imaginary line between the two. For other users in the vicinity, this excess energy degrades the radio channel, or makes it completely unusable for conversations. As the user density increases, radio resources are in consequence exhausted eventually. Systems with higher spectral efficiency provide more data *throughput* (services) with a given amount of spectrum and support more users at a given grade of service before experiencing resource exhaustion. The key benefits of higher spectral efficiencies can be enumerated as follows: higher aggregate capacity (per-cell throughput); higher per-user quality and service levels; higher subscriber density per base station; small spectrum requirements; and lower capital and operational costs in deployment. The spectral efficiency for various systems can be calculated easily using

$$\text{Spectral Efficiency} = \frac{\text{Channel Throughput}}{\text{Channel Bandwidth}}. \qquad (2.1)$$

This simply sums the *throughput* over a channel in an operating network and divides by the channel bandwidth. To understand spectral efficiency calculations, consider the PCS-1900 (GSM) system which can be parameterized as follows: 200 KHz carriers, 8 time slots per carrier, 13.3 Kbps of user data per slot, effective reuse of 7 (i.e., effectively 7 channel groups at 100 percent network load, or only 1/7th of each channels throughput available per cell). The spectral efficiency is therefore:

$$SE = 8 \text{ slots} \times 13.3 \frac{\text{Kbps}}{\text{slot}}/200 \text{ KHz}/7 \text{ cells}$$
$$= 0.076 \text{ b/s/Hz/cell}. \tag{2.2}$$

This value of approximately 0.1 b/s/Hz/cell is generally representative of high-mobility 2G and 3G cellular systems, including CDMA systems of all types. It reflects the fact that the classical techniques for increasing spectral efficiency have been exhausted and that new techniques are necessary [45]. Finally, it should be noted that the value of approximately 0.1 b/s/Hz/cell represents a major stumbling block for the delivery of next-generation services. Without substantial increases in spectral efficiency, 3G systems are bound to spectral efficiencies like those of todays 2G systems. In a typical 3G system with a 5 MHz downlink channel block, this translates into a total cell capacity of approximately 500 Kbps for the entire cell. With services advertised in the range of 144–384 Kbps, 1–3 users will completely occupy the cell capacity! This is far from the approximately 250–500 subscribers per cell needed to make the system economically viable, and it underscores the need for new methods to boost spectral efficiency.

2.7   MULTIPLE ACCESS SCHEMES

Mobile communications utilize the range of available frequencies in a number of ways, referred to as multiple-access schemes. Some basic schemes are FDMA, TDMA, CDMA, and OFDM.

2.7.1   FDMA

In the standard analog frequency division multiple access (FDMA) systems, such as AMPS, the most widely cellular phone system installed in North America, different carrier frequencies are allocated to different users. Individual conversations use communication channels appropriately separated in the frequency domain. In a system using the FDMA scheme, six frequencies are assigned to six users, and six simultaneous calls may be made as shown in Fig. 2.10(a). FDMA systems transmit one voice circuit per channel. Each conversation gets its own, unique, radio channel. The channels are relatively narrow, usually 30 KHz or less, and are defined as either transmit or receive channels. A full duplex conversation requires a transmit and receive

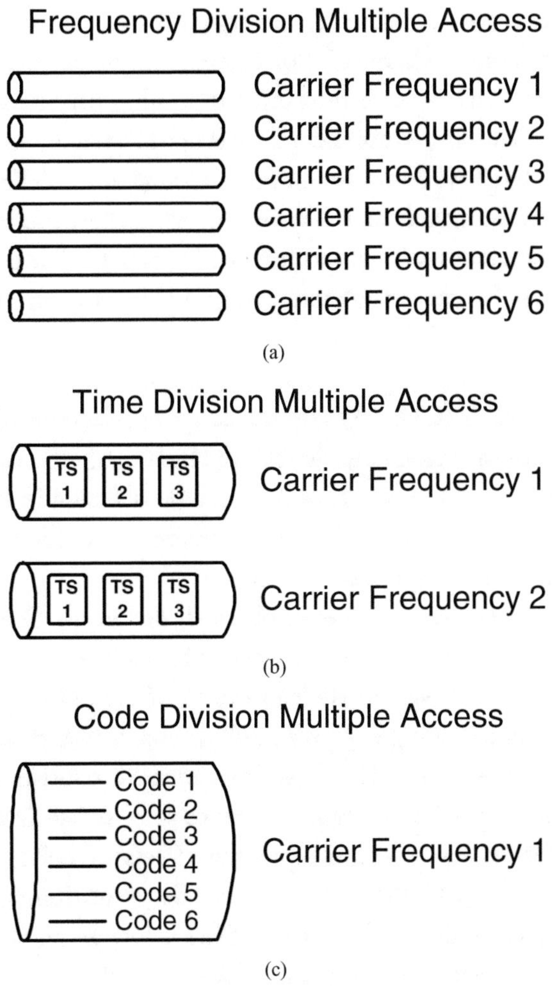

Frequency Division Multiple Access

```
(                    )  Carrier Frequency 1
(                    )  Carrier Frequency 2
(                    )  Carrier Frequency 3
(                    )  Carrier Frequency 4
(                    )  Carrier Frequency 5
(                    )  Carrier Frequency 6
```

(a)

Time Division Multiple Access

```
( [TS 1] [TS 2] [TS 3] )  Carrier Frequency 1

( [TS 1] [TS 2] [TS 3] )  Carrier Frequency 2
```

(b)

Code Division Multiple Access

```
( — Code 1
  — Code 2
  — Code 3    Carrier Frequency 1
  — Code 4
  — Code 5
  — Code 6 )
```

(c)

FIGURE 2.10: Channel usage for different multiple access schemes: (a) FDMA; (b) TDMA; (c) CDMA [40].

channel pair. For example, if a FDMA system had 200 channels, the system can handle 100 simultaneously full duplex conversations (100 channels for transmitting and 100 channels for receiving).

2.7.2 TDMA

With time division multiple access (TDMA) systems, separate conversations in both frequency and time domains take place, as shown in Fig. 2.10(b). Each frequency (channel) supports multiple conversations, which use the channel during specific time slots. Typically there is a maximum number of conversations which can be supported on each physical channel and each conversation occupies a logical "channel." For example, a system using this scheme creates two TDMA channels and divides each into three time slots, serving six users. *Global System Mobile* (GSM) communications, a unified pan-European system, is a time division-based digital cellular system. It employs 8 user time slots per frame in a 200 KHz channel. Like other TDMA systems, staggered transmit and receive time slots allow modems to use half-duplex radios, thereby reducing their costs. The transmit/receive offset still leaves enough idle time for the mobile to participate in handovers by monitoring neighboring cell channel signal strengths.

2.7.3   CDMA

Code Division Multiple Access (CDMA) systems use spread-spectrum (SS) signaling to create wideband sequences for transmission. This is achieved by several methods, such as pseudonoise (PN) sequences, frequency- or time-hopping techniques, as shown in Fig. 2.10(c). A number of users simultaneously and asynchronously access a channel by modulating their information-bearing signals with preassigned signature sequences [51].

In the case of PN sequences, for example, also known as Direct Sequence CDMA (DS-CDMA), each user in the system uses a separate code for transmission, as shown in Fig. 2.10(c). The design aims to spread the bandwidth of the information sequence by multiplying it by a PN sequence yielding a longer random sequence and simultaneously reducing the spectral density of the signal [40]. This new sequence consists of inverted and non-inverted versions of the original PN sequence. Since it is noisy-like, it possesses a wider frequency bandwidth that is less susceptible to the effects of noise and narrowband jammers during transmission [52]. CDMA systems provide protection against multipath interference and antijamming capability. Additionally, there is low probability of interception and, thus, unauthorized parties become less capable of detecting the information message during transmission.

In frequency hopping CDMA (FH-CDMA), each user is identified by a unique spreading sequence to create a pseudo random hop pattern of the transmission frequencies over the entire bandwidth. These sequences are available at the receiver to identify the users. In frequency hopping CDMA, the carrier frequency of the modulated information signal is not constant but changes periodically. During time intervals T, the carrier frequency remains the same, but after each time interval the carrier hops to another (or possibly the same) frequency. The hopping pattern is decided by the spreading code. The set of available frequencies the carrier can attain is called the hop-set. The frequency occupation of an FH-SS system differs considerably from a DS-SS system. A DS system occupies the entire frequency band when it transmits, whereas an FH system uses only a small part of the bandwidth when it transmits, but the location of this part differs in time.

In time-hopping CDMA (TH-CDMA), the information-bearing signal is not transmitted continuously. Instead, the signal is transmitted in short bursts at time intervals determined by the spreading code assigned to the user. In-time hopping CDMA the data signal is transmitted in rapid bursts at time intervals. The time axis is divided into frames, and each frame is divided into M, for example, time slots. During each frame the user transmits in one of the M time slots. The code signal assigned to the user defines which of the M time slots is transmitted. Since a user transmits all of its data in one, instead of M time slots, the frequency it needs for its transmission increases by a factor of M.

In theory, the capacity provided by the three multiple access schemes is the same and is not altered by dividing the spectrum into frequencies, time slots, or codes, as explained in the following example [53]. Assume that there are six carrier frequencies available for transmission covering the available bandwidth. The channel usage for FDMA, TDMA, and CDMA is depicted in Fig. 2.10. In a system using the FDMA scheme, six frequencies are assigned to six users, and six simultaneous calls may be made. TDMA generally requires a larger bandwidth than FDMA. A system using this scheme can create two TDMA channels and divides each into three time slots, serving six users [40]. A CDMA channel requires a larger bandwidth than the other two and serves six calls by using six codes, as illustrated in Fig. 2.10(c).

2.7.4   OFDM

The principle of orthogonal frequency division multiple (OFDM) access has existed for several decades. However, it was only in the last decade that it started to be used in commercial systems. Digital Audio and Video Broadcasting (DAB and DVB), wireless local area networks (WLAN), and more recently wireless local loop (WLL) are the most important wireless applications that use OFDM [54]. The main concept of the method is that one data stream, of Q bps for example, is divided into N data streams, each at a rate of Q/N bps where each one is carried by a different frequency. In OFDM, the subcarrier pulse used for transmission is chosen to be rectangular. This has the advantage that the task of pulse forming and modulation can be performed by a simple Inverse Discrete Fourier Transform (IDFT). Thus, the N data streams are combined together using the Inverse Fast Fourier Transform (IFFT), which can be implemented very efficiently, to obtain a time-domain waveform for transmission as an IFFT. Therefore, in the receiver, a forward FFT is needed to reverse this operation. According to the theorems of the Fourier Transform the rectangular pulse shape will lead to a $\sin(x)/x$ spectrum of the subcarriers as shown in Fig. 2.11.

The parallel, and slower data streams, are allowed to overlap in frequency. In this way, the bandwidth of the modulated symbol effectively decreases by N, and its duration increases by N, as well. Therefore, with the appropriate choice of N, frequency-selectivity and ISI (Inter

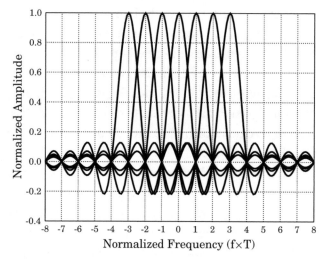

FIGURE 2.11: OFDM and the orthogonality principle.

Symbol Interference) can be removed. The carrier frequency spacing Δf is selected so that each subcarrier is orthogonal to all other subcarriers, thus $\Delta f = 1/T$, where T is the OFDM symbol duration (or, more precisely, the effective duration of the Fourier transform).

OFDM is particularly suited for transmission over a dispersive (i.e., frequency selective) channel. In 1993 Linnertz et al. proposed the multi-carrier code division multiple access (MC-CDMA) [55]. It is a new CDMA system based on a combination of CDMA and orthogonal frequency division OFDM where the spreading is performed in the frequency domain, rather than in the time domain as in a DS-CDMA system. In MC-CDMA, each of the M carriers in a multi-carrier system is multiplied by a spreading sequence unique to each user. This system has gained much attention, because the signal can be easily transmitted and received using the Fast Fourier Transform (FFT) device without increasing the transmitter and receiver complexities and is potentially robust to channel frequency selectivity with a good frequency use efficiency [56].

CHAPTER 3

Antenna Arrays and Diversity Techniques

An antenna in a telecommunications system is the device through which, in the transmission mode, *radio frequency* (RF) energy is coupled from the transmitter to the free space, and from free space to the receiver in the receiving mode [57–59].

3.1 ANTENNA ARRAYS

In many applications, it is necessary to design antennas with very directive characteristics (very high gains) to meet demands for long distance communication. In general, this can only be accomplished by increasing the electrical size of the antenna. Another effective way is to form an assembly of radiating elements in a geometrical and electrical configuration, without necessarily increasing the size of the individual elements [9]. Such a multielement radiation device is defined as an *antenna array* [59].

The total electromagnetic field of an array is determined by vector addition of the fields radiated by the individual elements, combined properly in both amplitude and phase[58, 59]. Antenna arrays can be one-, two-, and three-dimensional. By using basic array geometries, the analysis and synthesis of their radiation characteristics can be simplified. In an array of identical elements, there are at least five individual controls (degrees of freedom) that can be used to shape the overall pattern of the antenna. These are the [59]:

i.   geometrical configuration of the overall array (linear, circular, rectangular, spherical, etc.)

ii. relative displacement between the elements

iii. amplitude excitation of the individual elements

iv. phase excitation of the individual elements

v.   relative pattern of the individual elements

3.2   ANTENNA CLASSIFICATION

In general, antennas of individual elements may be classified as *isotropic, omnidirectional* and *directional* according to their radiation characteristics. Antenna arrays may be referred to as *phased arrays* and *adaptive arrays* according to their functionality and operation [59].

3.2.1   Isotropic Radiators

An isotropic radiator is one which radiates its energy equally in all directions. Even though such elements are not physically realizable, they are often used as references to compare to them the radiation characteristics of actual antennas.

3.2.2   Omnidirectional Antennas

Omnidirectional antennas are radiators having essentially an isotropic pattern in a given plane (the azimuth plane in Fig. 3.1) and directional in an orthogonal plane (the elevation plane in Fig. 3.1). Omnidirectional antennas are adequate for simple RF environments where no specific knowledge of the users directions is either available or needed. However, this unfocused approach scatters signals, reaching desired users with only a small percentage of the overall energy sent out into the environment [4]. Thus, there is a waste of resources using omnidirectional antennas since the vast majority of transmitted signal power radiates in directions other than the desired user. Given this limitation, omnidirectional strategies attempt to overcome environmental challenges by simply increasing the broadcasting power. Also, in a setting of numerous users (and interferers), this makes a bad situation worse in that the signals that miss the intended user become interference for those in the same or adjoining cells. Moreover, the single-element approach cannot selectively reject signals interfering with those of served users. Therefore, it has no spatial multipath mitigation or equalization capabilities. Omnidirectional strategies directly and adversely impact spectral efficiency, limiting frequency reuse. These

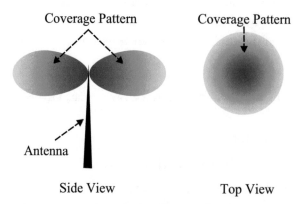

FIGURE 3.1: Omnidirectional antennas and coverage patterns [4].

limitations of broadcast antenna technology regarding the quality, capacity, and geographic coverage of wireless systems initiated an evolution in the fundamental design and role of the antenna in a wireless system.

3.2.3   Directional Antennas

Unlike an omnidirectional antenna, where the power is radiated equally in all directions in the horizontal (azimuth) plane as shown in Fig. 3.1, a directional antenna concentrates the power primarily in certain directions or angular regions [59]. The radiating properties of these antennas are described by a *radiation pattern*, which is a plot of the radiated energy from the antenna measured at various angles at a constant radial distance from the antenna. In the near field the relative radiation pattern (shape) varies accorging to the distance from the antenna, whereas in the far field the relative radiation pattern (shape) is basically independent of distance from the antenna. The direction in which the intensity/gain of these antennas is maximum is referred to as the *boresight* direction [59, 60]. The gain of directional antennas in the boresight direction is usually much greater than that of isotropic and/or omnidirectional antennas. The radiation pattern of a directional antenna is shown in Fig. 3.2 where the boresight is in the direction $\theta = 0°$. The plot consists of a main lobe (also referred to as *major lobe*), which contains the boresight and several minor lobes including side and rear lobes. Between these lobes are directions in which little or no radiation occurs. These are termed *minima* or *nulls*. Ideally, the intensity of the field toward nulls should be zero (minus infinite *dBs*). However, practically nulls may represent a 30 or more dB reduction from the power at boresight. The angular segment subtended by two points where the power is one-half the main lobe's peak value is known as the *half-power beamwidth*.

3.2.4   Phased Array Antennas

A phased array antenna uses an array of single elements and combines the signal induced on each element to form the array output. The direction where the maximum gain occurs is usually controlled by adjusting properly the amplitude and phase between the different elements [59]. Fig. 3.3 describes the phased array concept.

3.2.5   Adaptive Arrays

Adaptive arrays for communication have been widely examined over the last few decades. The main thrust of these efforts has been to develop arrays that would provide both interference protection and reliable signal acquisition and tracking in communication systems [61]. The radiation characteristics of these arrays are adaptively changing according to changes and requirements of the radiation environment. Research on adaptive arrays has involved both theoretical and experimental studies for a variety of applications. The field of adaptive array

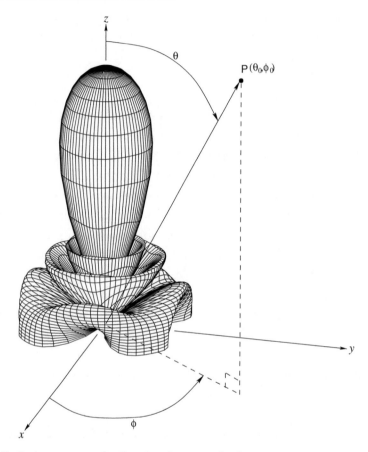

FIGURE 3.2: Radiation pattern of a directional antenna [17].

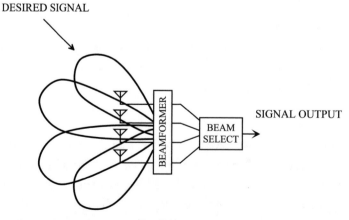

FIGURE 3.3: Phased array antenna concept [20].

sensor systems has now become a mature technology, and there is a wealth of literature available on various aspects of such systems [62].

Adaptive arrays provide significant advantages over conventional arrays in both communication and radar systems. They have well-known advantages for providing flexible, rapidly configurable, beamforming and null-steering patterns [62]. However, this is often assumed because of its flexibility in using the available array elements in an adaptive mode and, thus, can overcome most, if not all, of the deficiencies in the design of the basic or conventional arrays [63]. Therefore, conventional goals, such as low sidelobes and narrow beamwidth in the array design can be ignored in the implementation of an adaptive array. Nevertheless, much work has drawn attention toward these impairments of adaptive arrays and reported the serious problems, such as grating nulls, with improper selection of element distributions and patterns [64].

An adaptive antenna array is the one that continuously adjusts its own pattern by means of feedback control [9]. The principal purpose of an adaptive array sensor system is to enhance the detection and reception of certain desired signals [62]. The pattern of the array can be steered toward a desired direction space by applying phase weighting across the array and can be shaped by amplitude and phase weighting the outputs of the array elements [65]. Additionally, adaptive arrays sense the interference sources from the environment and suppress them automatically, improving the performance of a radar system, for example, without *a priori* information of the interference location [66]. In comparison with conventional arrays, adaptive arrays are usually more versatile and reliable.

A major reason for the progress in adaptive arrays is their ability to automatically respond to an unknown interfering environment by steering nulls and reducing side lobe levels in the direction of the interference, while keeping desired signal beam characteristics [66]. Most arrays are built with fixed weights designed to produce a pattern that is a compromise between resolution, gain, and low sidelobes. However, the versatility of the array antenna invites the use of more sophisticated techniques for array weighting [65]. Particularly attractive are adaptive schemes that can sense and respond to a time-varying environment. The precise control of null placement in adaptive arrays results in slight deterioration in the output SNR.

Adaptive antenna arrays are commonly equipped with signal processors which can automatically adjust by a simple adaptive technique the variable antenna weights of a signal processor so as to maximize the signal-to-noise ratio. At the receiver output, the desired signal along with interference and noise are received at the same time. The adaptive antenna scans its radiation pattern until it is fixed to the optimum direction (toward which the signal-to-noise ratio is maximized). In this direction the maximum of the pattern is ideally toward the desired signal.

Adaptive arrays based on DSP algorithms can, in principle, receive desired signals from any angle of arrival. However, the output signal-to-interference plus-noise ratio (SINR) obtained from the array, as the desired and interference signal angles of arrival and polarizations vary, depends critically on the element patterns and spacings used in the array [61].

3.3   DIVERSITY TECHNIQUES

Diversity combining [67] is an effective way to overcome the problem of fading in radio channels. It utilizes the fact that if some receive antennas are experiencing a low signal level due to fading, also called a deep fade, some others will probably not suffer from the same deep fade, provided that they are displaced in appropriate positions, or in polarity [68].

Let us now consider the transmission of an information sequence over a frequency non-selective channel. The average bit error probability (BEP) is given by

$$P_b = \int_0^\infty P_b(\gamma_b) p(\gamma_b) \mathrm{d}\gamma_b \tag{3.1}$$

where $P_b(\gamma_b)$ is the bit error probability as a function of the received signal-to-noise-ratio (SNR), γ_b, and $p(\gamma_b)$ is the probability density function (PDF) of the received SNR. As an example, we examine the transmission of Binary Phase Shift Keying (BPSK) information sequence over a Rayleigh fading channel. In this case, $P_b(\gamma_b)$ is given by

$$P_b(\gamma_b) = Q\left(\sqrt{\gamma_b}\right) \tag{3.2}$$

where $\gamma_b = \alpha^2 E_b / N_0$ is the received SNR and E_b is the energy of the transmitted information bit. Moreover, for a Rayleigh fading channel, it can be easily shown that

$$p(\gamma_b) = \frac{1}{\bar{\gamma}_b} e^{-\gamma_b/\bar{\gamma}_b} \tag{3.3}$$

where $\bar{\gamma}_b$ is the average SNR defined by

$$\bar{\gamma}_b = \frac{E_b}{N_0} \mathcal{E}\left\{\alpha^2\right\} \tag{3.4}$$

where $\mathcal{E}\{\cdot\}$ denotes the expectation value. Substituting $P_b(\gamma_b)$ and $p(\gamma_b)$ into the expression for P_b in (3.1), we obtain the average bit error probability as

$$P_b = \frac{1}{2}\left(1 - \sqrt{\frac{\bar{\gamma}_b}{1 + \bar{\gamma}_b}}\right). \tag{3.5}$$

The bit error probabilities for BPSK modulation over AWGN and Rayleigh fading channels are shown in Fig. 3.4. When simulating the performance of any information bearing sequence

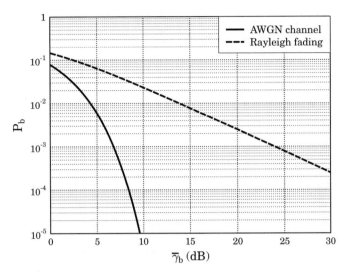

FIGURE 3.4: Bit error probability for BPSK modulation over AWGN and Rayleigh fading channels.

transmitted through a particular wireless channel, a bit error occurs if the decision for the received bit does not match the originally transmitted data bit. The bit error rate (BER) is the ratio of the number of bit errors to the total number of transmitted data bits [69]. From Fig. 3.4, we observe that while the error probability decreases exponentially with SNR for the AWGN channel, it decreases only inversely for the Rayleigh fading channel case [70]. Therefore, fading degrades the performance of a wireless communication system significantly.

In order to combat fading, the receiver is typically provided with multiple replicas of the transmitted signal. In this way, the transmitted information is extracted with the minimum possible number of errors since all the replicas do not typically fade simultaneously. This method is called *diversity* and is one of the most effective techniques to combat multipath fading. There exist many diversity techniques including temporal, frequency, space, and polarization diversity. A block diagram of a digital communication system with diversity is shown in Fig. 3.5. The diversity combiner combines the received signals from the different diversity branches. The combiner simply exploits the information embedded in each branch to form the decision variable [26, 70].

In temporal diversity, the same signal is transmitted at different times, where the separation between the time intervals is at least equal to the coherence time, T_c. Therefore, the separated in time channels fade independently and thus, proper diversity reception is achieved.

Frequency diversity exploits the fact that frequencies separated by at least the coherence bandwidth of the channel, B_c, fade almost independently of each other. Thus, if a signal is

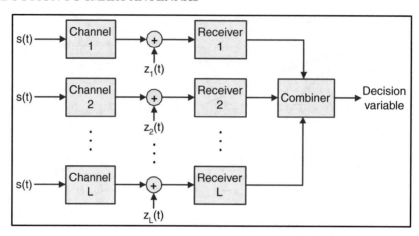

FIGURE 3.5: Model of a digital communication system with diversity [70].

transmitted simultaneously using frequencies appropriately apart from each other, the receiver is provided with independent fading branches through several frequency channels.

In spatial (antenna) diversity, spatially separated antennas are used at the transmitter and/or the receiver. In this way, the replicas of the transmitted signal are provided to the receiver via separate spatial channels [26, 70]. It has been shown that a spatial separation of at least half-wavelength is necessary that the signals received from antenna elements are (almost) independent in a rich scattering, or more precisely in a uniform scattering environment [71].

In antenna diversity, signals received by the different antenna branches are demodulated to baseband with quadrature demodulator and processed with correlator or matched filter detector. The output is then applied to a diversity combiner. This procedure guarantees that fading will be slow and generally not change through a time slot. The option to select the best antenna significantly improves performance [68].

One method of combining in spatial diversity is to weight each diversity branch with its complex conjugate of its own channel gain (so that the phase introduced by the channel to be as much as possible removed). The combiner then adds the outputs of this process from each individual branch to form its decision. This technique is also known as the *maximum ratio combiner* (MRC) and is the optimal diversity scheme. However, it needs perfect channel knowledge for maximum performance.

Although optimal, MRC is expensive to implement and requires an accurate tracking of the complex fading which is difficult to achieve in practice [26]. *Equal gain combining* (EGC) diversity technique is a simple alternative to MRC. It consists of the co-phasing of the signals received from each diversity branch using unit weights before added by the combiner [26]. The performance of EGC is found to be very close to that of MRC. The SNR of the combined signals using EGC is only 1 dB below the SNR provided by MRC [72].

In *switched diversity* (SC), the decision is made using a branch with SNR larger than a predetermined threshold. If the SNR drops below this threshold, the combiner switches to another branch that satisfies the threshold criterion.

Another combining scheme is *selection diversity* (SC) in which all the branches are monitored simultaneously [70]. The branch yielding the highest SNR ratio is always selected at any one time. The received signal is then multiplied by the complex conjugate of the corresponding branch. The formed decision is based upon this output.

At this point, it would be useful to see the performance of a particular antenna diversity scheme. For example, employing MRC and BPSK modulation, the probability of bit error is given by

$$P_b = \left(\frac{1-\mu}{2}\right)^L \sum_{l=0}^{L-1} \binom{L-1+l}{l} \left(\frac{1+\mu}{2}\right)^l \tag{3.6}$$

where L is the number of the present diversity branches and $\mu = \sqrt{\frac{\overline{\gamma}_b}{1+\overline{\gamma}_b}}$. For large values of the average bit-to-noise ratio, (3.6) simplifies to [73]

$$P_b \approx \left(\frac{1}{4\overline{\gamma}_b}\right) \binom{2L-1}{L}. \tag{3.7}$$

Thus, at high values of SNR, it possesses in its diagram a slope approximately equal to $-L$ dB/decade. Fig. 3.6 shows the performance of MRC for different number of branches L. As the diversity order increases, the BER performance is improved, or equivalently there is a significant gain in SNR for a given BER. However, this increase in performance is accompanied by the trade-off of more expensive and complicated infrastructure and additional required transmission power.

The polarization diversity scheme achieves its diversity based on the different propagation characteristics of the vertically and horizontally polarized electromagnetic waves [74]. Polarization diversity is different from space diversity. It is based on the concept that in high multipath environments, the signal from a portable received at the base station has varying polarization. The mechanism of decorrelation for the different polarizations is the multipath reflections encountered by a signal traveling between the portable and base station. Typically, an improvement in the uplink performance can be achieved by using two receive antennas with orthogonal polarizations and combining these signals. Because the two receive antennas do not need to be spaced apart horizontally to accomplish this, they can be mounted under the same radome [75]. Polarization diversity does have its benefits. It is easy to obtain a suitable site

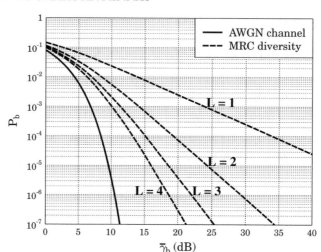

FIGURE 3.6: The MRC diversity technique [72].

because large structures that are required for space-diversity techniques are not needed. But polarization diversity is completely effective only in high multipath environments. Some manufacturers have promoted polarization diversity as performing better than space diversity in all environments [75]. However, when high multipath environments do not exist, the performance of the polarization-diversity antennas may not be as good as the space-diversity system. Polarization diversity is a useful technique in the proper environment, where the necessary multipath is present. Before assuming that polarization diversity may work in a particular environment, field testing must be performed to compare space diversity and polarization diversity.

Angular antenna diversity has been considered as an attempt to control the dispersive type of fading along with the traditional antenna space diversity being utilized to reduce the impact of flat fading [76]. In angle diversity, antennas with narrow beamwidths are positioned in different angular directions or regions. The use of narrower beams increases the gain of the base station antenna and provides angular discrimination that can reduce interference [77]. Furthermore, it has been shown practically, by Perini [77] and others, that the effect of angular diversity is quite similar to that of using space diversity, especially in dense urban areas.

Fig. 3.7 shows three antenna diversity options with four antenna elements for a 120° sectorized system. Fig. 3.7(a) shows spatial diversity with approximately seven wavelengths (7λ) spacing between the elements (3.3 m at 1900 MHz). A typical antenna element has a gain of 18 dBi. The horizontal and vertical beamwidths are 65° and 80°, respectively. Fig. 3.7(b) shows two dual polarization antennas, where the antennas can be either closely spaced (λ/2) to provide both angle and polarization diversity in a small profile, or widely spaced (7λ) to provide both spatial and polarization diversity [20]. The antenna elements shown are the commonly

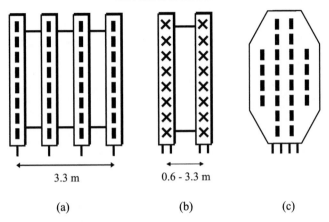

FIGURE 3.7: Antenna diversity options with four antenna elements: (a) spatial diversity; (b) polarization diversity with angular and spatial diversity; (c) angular diversity [20].

used 45° slant polarization antennas, rather than vertically and horizontally polarized antennas. Finally, in Fig. 3.7(c) a closely spaced ($\lambda/2$) vertically polarized array is shown. Such an array provides angle diversity in a small profile [20].

CHAPTER 4

Smart Antennas

4.1 INTRODUCTION

Many refer to smart antenna systems as smart antennas, but in reality antennas by themselves are not smart. It is the digital signal processing capability, along with the antennas, which make the system smart. Although it may seem that smart antenna systems are a new technology, the fundamental principles upon which they are based are not new. In fact, in the 1970s and 1980s two special issues of the *IEEE Transactions on Antennas and Propagation* were devoted to adaptive antenna arrays and associated signal processing techniques [78, 79]. The use of adaptive antennas in communication systems initially attracted interest in military applications [27]. Particularly, the techniques have been used for many years in electronic warfare (EWF) as countermeasures to electronic jamming. In military radar systems, similar techniques were already used during World War II [80]. However, it is only because of todays advancement in powerful low-cost digital signal processors, general-purpose processors and ASICs (Application Specific Integrated Circuits), as well as innovative software-based signal processing techniques (algorithms), that smart antenna systems are gradually becoming commercially available [17, 59].

4.2 NEED FOR SMART ANTENNAS

Wireless communication systems, as opposed to their wireline counterparts, pose some unique challenges [42]:

  i.   the limited allocated spectrum results in a limit on capacity

 ii.   the radio propagation environment and the mobility of users give rise to signal fading and spreading in time, space and frequency

iii.   the limited battery life at the mobile device poses power constraints

In addition, cellular wireless communication systems have to cope with interference due to frequency reuse. Research efforts investigating effective technologies to mitigate such effects have been going on for the past twenty five years, as wireless communications are experiencing rapid growth [42]. Among these methods are multiple access schemes, channel coding and

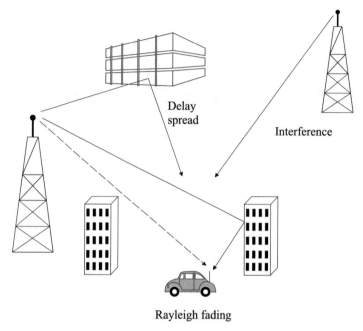

FIGURE 4.1: Wireless systems impairments [81].

equalization and smart antenna employment. Fig. 4.1 summarizes the wireless communication systems impairments that smart antennas are challenged to combat.

An antenna in a telecommunications system is the port through which *radio frequency* (RF) energy is coupled from the transmitter to the outside world for transmission purposes, and in reverse, to the receiver from the outside world for reception purposes [57, 59]. To date, antennas have been the most neglected of all the components in personal communications systems. Yet, the manner in which radio frequency energy is distributed into and collected from space has a profound influence upon the efficient use of spectrum, the cost of establishing new personal communications networks and the service quality provided by those networks [20]. The commercial adoption of smart antenna techniques is a great promise to the solution of the aforementioned wireless communications' impairments.

4.3   OVERVIEW

The basic idea on which smart antenna systems were developed is most often introduced with a simple intuitive example that correlates their operation with that of the human auditory system. A person is able to determine the Direction of Arrival (DoA) of a sound by utilizing a three-stage process:

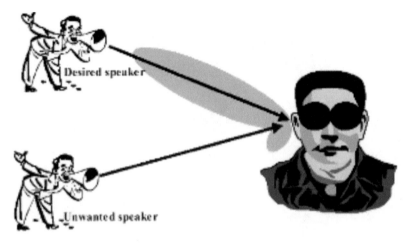

FIGURE 4.2: Human auditory function [17].

- One's ears act as acoustic sensors and receive the signal.
- Because of the separation between the ears, each ear receives the signal with a different time delay.
- The human brain, a specialized signal processor, does a large number of calculations to correlate information and compute the location of the received sound.

To better provide an insight of how a smart antenna system works, let us imagine two persons carrying on a conversation inside an isolated room as illustrated in Fig. 4.2. The listener among the two persons is capable of determining the location of the speaker as he moves about the room because the voice of the speaker arrives at each acoustic sensor, the ear, at a different time. The human "signal processor," the brain, computes the direction of the speaker from the time differences or delays received by the two ears. Afterward, the brain adds the strength of the signals from each ear so as to focus on the sound of the computed direction.

Utilizing a similar process, the human brain is capable of distinguishing between multiple signals that have different directions of arrival. Thus, if additional speakers join the conversation, the brain is able to enhance the received signal from the speaker of interest and tune out unwanted interferers. Therefore, the listener has the ability to distinguish one person's voice, from among many people talking simultaneously, and concentrate on one conversation at a time. In this way, any unwanted interference is attenuated. Conversely, the listener can respond back to the same direction of the desired speaker by orienting his/her transmitter, his/her mouth, toward the speaker.

Electrical smart antenna systems work the same way using two antennas instead of two ears, and a digital signal processor instead of the brain as seen in Fig. 4.3. Thus, based on the

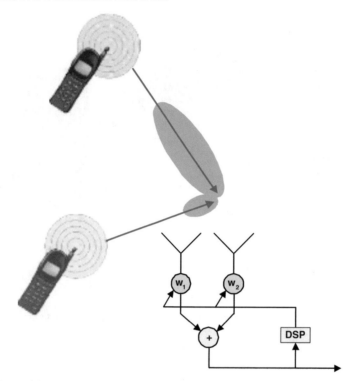

FIGURE 4.3: A two-element electrical smart antenna.

time delays due to the impinging signals onto the antenna elements, the digital signal processor computes the direction-of-arrival (DOA) of the signal-of-interest (SOI), and then it adjusts the excitations (gains and phases of the signals) to produce a radiation pattern that focuses on the SOI while tuning out any interferers or signals-not-of-interest (SNOI).

Transferring the same idea to mobile communication systems, the base station plays the role of the listener, and the active cellular telephones simulate the role of the several sounds heard by human ears. The principle of a smart antenna system is illustrated in Fig. 4.4.

A digital signal processor located at the base station works in conjunction with the antenna array and is responsible for adjusting various system parameters to filter out any interferers or *signals-not-of-interest* (SNOI) while enhancing desired communication or *signals-of-interest* (SOI). Thus, the system forms the radiation pattern in an adaptive manner, responding dynamically to the signal environment and its alterations. The principle of beamforming is essentially to weight the transmit signals in such a way that the receiver obtains a constructive super-position of different signal parts. Note that some knowledge of the transmission channel at the transmitter is necessary in order for beamforming to be feasible [82]. A comprehensive overview of beamforming techniques is given in [83]. Fig. 4.5 illustrates the general idea of adaptive beamforming.

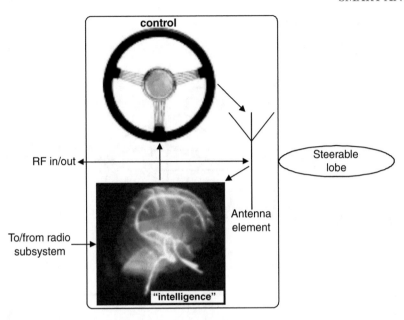

FIGURE 4.4: Principle of a smart antenna system [80].

4.4   SMART ANTENNA CONFIGURATIONS

Basically, there are two major configurations of smart antennas:

- Switched-Beam: A finite number of fixed, predefined patterns or combining strategies (sectors).

- Adaptive Array: A theoretically infinite number of patterns (scenario-based) that are adjusted in real time according to the spatial changes of SOIs and SNOIs.

In the presence of a low level interference, both types of smart antennas provide significant gains over the conventional sectorized systems. However, when a high level interference is present, the interference rejection capability of the adaptive systems provides significantly more coverage than either the conventional or switched beam system [4]. Fig. 4.6 illustrates the relative coverage area for conventional sectorized, switched-beam, and adaptive antenna systems.

Both types of smart antenna systems provide significant gains over conventional sectorized systems. The low level of interference environment on the left represents a new wireless system with lower penetration levels. However the environment with a significant level of interference on the right represents either a wireless system with more users or one using more aggressive frequency reuse patterns. In this scenario, the interference rejection capability of the adaptive system provides significantly more coverage than either the conventional or switched beam systems [4].

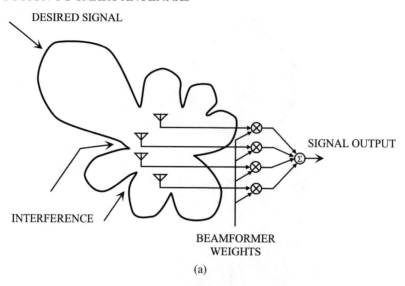

(a)

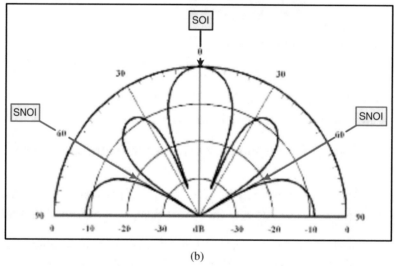

(b)

FIGURE 4.5: Adaptation procedure: (a) Calculation of the beamformer weights [20] and (b) Beamformed antenna amplitude pattern to enhance SOI and suppress SNOIs.

Now, let us assume that a signal of interest and two co-channel interferers arrive at the base station of a communications system employing smart antennas. Fig. 4.7 illustrates the beam patterns that each configuration may form to adapt to this scenario.

The switched-beam system is shown on the left while the adaptive system is shown on the right. The light lines indicate the signal of interest while the dark lines display the direction of the co-channel interfering signals. Both systems direct the lobe with the greatest intensity in the general direction of the signal of interest. However, switched fixed beams achieve coarser pattern

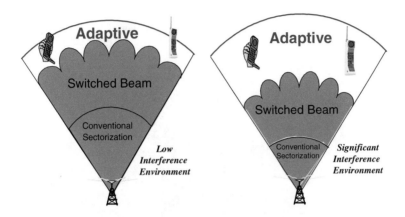

FIGURE 4.6: Coverage patterns for switched beam and adaptive array antennas [20].

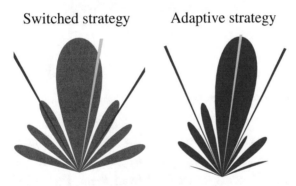

FIGURE 4.7: Beamforming lobes and nulls that Switched-Beam (left) and Adaptive Array (right) systems might choose for identical user signals (light line) and co-channel interferers (dark lines) [20].

control than adaptive arrays [84]. The adaptive system chooses a more accurate placement, thus providing greater signal enhancement. Similarly, the interfering signals arrive at places of lower intensity outside the main lobe, but again the adaptive system places these signals at the lowest possible gain points. The adaptive array concept ideally ensures that the main signal receives maximum enhancement while the interfering signals receive maximum suppression.

4.4.1 Switched-Beam Antennas

A switched-beam system is the simplest smart antenna technique. It forms multiple fixed beams with heightened sensitivity in particular directions. Such an antenna system detects signal strength, chooses from one of several predetermined fixed beams, and switches from one beam to another as the cellular phone moves throughout the sector, as illustrated in Fig. 4.8.

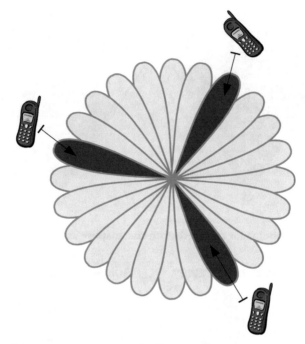

FIGURE 4.8: Switched-beam coverage pattern [85].

The switched-beam, which is based on a basic switching function, can select the beam that gives the strongest received signal. By changing the phase differences of the signals used to feed the antenna elements or received from them, the main beam can be driven in different directions throughout space. Instead of shaping the directional antenna pattern, the switched-beam systems combine the outputs of multiple antennas in such a way as to form narrow sectorized (directional) beams with more spatial selectivity that can be achieved with conventional, single-element approaches. Other sources in the literature [86] define this concept as *phased array* or *multibeam antenna*. Such a configuration consists of either a number of fixed beams with one beam turned on toward the desired signal or a single beam (formed by phase adjustment only) that is steered toward the desired signal.

A more generalized to the Switched-Lobe concept is the Dynamical Phased Array (DPA). In this concept, a direction of arrival (DOA) algorithm is embedded in the system [20]. The DOA is first estimated and then different parameters in the system are adjusted in accordance with the desired steering angle. In this way the received power is maximized but with the trade-off of more complicated antenna designs.

The elements used in these arrays must be connected to the sources and/or receivers by feed networks. One of the most widely-known multiple beamforming networks is the *Butler matrix* [87, 88]. It is a linear, passive feeding, $N \times N$ network with beam steering capabilities

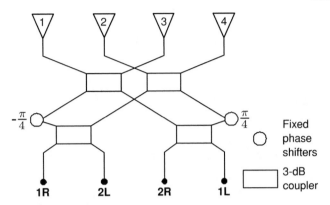

FIGURE 4.9: A schematic diagram of a 4 × 4 Butler matrix [90].

for phased array antennas with N outputs connected to antenna elements and N inputs or beam ports. The Butler matrix performs a spatial fast Fourier transform and provides N orthogonal beams, where N should be an integer power of 2 (i.e. $N = 2^n$, $n \in \mathbb{Z}^+$) [89]. These beams are linear independent combinations of the array element patterns. A Butler matrix-fed array can cover a sector of up to 360° depending on element patterns and spacing. Each beam can be used by a dedicated transmitter and/or receiver and the appropriate beam can be selected using an RF switch. A Butler matrix can also be used to steer the beam of a circular array by exciting the Butler matrix beam ports with amplitude and phase weighted inputs followed by a variable uniform phase taper [89]. The only required transmit/receive chain combines alternate rows of hybrid junctions (or directional couplers) and fixed phase shifters [90]. Fig. 4.9 shows a schematic diagram of a 4 × 4 Butler matrix.

A total of $(N/2) \times \log_2 N$ hybrids and $(N/2) \times \log_2(N-1)$ fixed phase shifters are required to form the network. The hybrids can be either 90° or 180° 3 dB hybrids, depending on if the beams are to be symmetrical distributed about the broadside or whether one of the beams is to be in the broadside direction [91]. A Butler matrix serves two functions:

i.   distribution of RF signals to radiating antenna elements and

ii. orthogonal beam forming and beam steering.

By connecting a Butler matrix between an antenna array and an RF switch, multiple beam-forming can be achieved by exciting two or more beam ports with RF signals at the same time. A signal introduced at an input port will produce equal excitations at all output ports with a progressive phase between them, resulting in a beam radiated at a certain angle in space. A signal at another input port will form a beam in another direction, achieving beam steering. Referring to Fig. 4.10, if ports $1R$ and $4L$ are excited at the same time with RF signals of equal amplitude

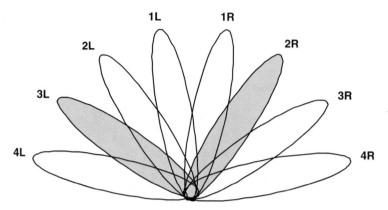

FIGURE 4.10: 8 orthogonal beams formed by an 8 × 8 Butler matrix [90].

and phase, beams $2R$ and $3L$ will radiate simultaneously. Although multiple beamforming is possible, there is a limitation. Two adjacent beams cannot be formed simultaneously as they will add to produce a single beam [92].

4.4.2 Adaptive Antenna Approach

The adaptive antenna systems approach communication between a user and a base station in a different way by adding the dimension of space. By adjusting to the RF environment as it changes (or the spatial origin of signals), adaptive antenna technology can dynamically alter the signal patterns to optimize the performance of the wireless system. *Adaptive array systems* [78, 79] provide more degrees of freedom since they have the ability to adapt in real time the radiation pattern to the RF signal environment; in other words, they can direct the main beam toward the pilot signal or SOI while suppressing the antenna pattern in the direction of the interferers or SNOIs. To put it simply, adaptive array systems can customize an appropriate radiation pattern for each individual user. Fig. 4.11 illustrates the general idea of an adaptive antenna system.

The adaptive concept is far superior to the performance of a switched-beam system, as it is shown in Fig. 4.6. Also, it shows that switched-beam system not only may not be able to place the desired signal at the maximum of the main lobe, but also it exhibits inability to fully reject the interferers. Because of the ability to control the overall radiation pattern in a greater coverage area for each cell site, as illustrated in Fig. 4.7, adaptive array systems can provide great increase in capacity. Adaptive array systems can locate and track signals (users and interferers) and dynamically adjust the antenna pattern to enhance reception while minimizing interference using signal processing algorithms. A functional block diagram of the digital signal processing part of an adaptive array antenna system is shown in Fig. 4.12.

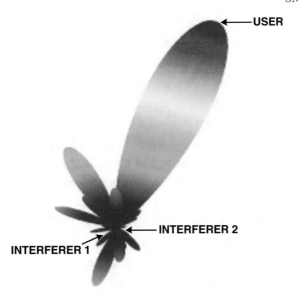

FIGURE 4.11: Adaptive array coverage: A representative depiction of a main lobe extending toward a user with nulls directed toward two co-channel interferers.

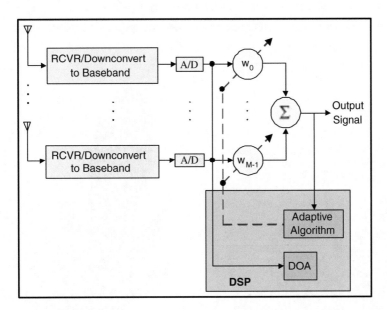

FIGURE 4.12: Functional block diagram of an adaptive array system.

After the system downconverts the received signals to baseband and digitizes them, it locates the SOI using the direction-of-arrival (DOA) algorithm, and it continuously tracks the SOI and SNOIs by dynamically changing the complex weights (amplitudes and phases of the antenna elements). Basically, the DOA computes the direction-of-arrival of all the signals by computing the time delays between the antenna elements, and afterward, the adaptive algorithm, using a cost function, computes the appropriate weights that result in an optimum radiation pattern. Because adaptive arrays are generally more digital processing intensive and require a complete RF portion of the transceiver behind each antenna element, they tend to be more expensive than switched-beam systems.

Adaptive arrays utilize sophisticated signal-processing algorithms to continuously distinguish between desired signals, multipath, and interfering signals, as well as calculate their Directions of Arrival (DOA). This approach updates its transmit strategy continuously based on changes in both the desired and interfering signal locations. A number of well-documented algorithms exist for estimating the DOA; for example, MUSIC, ESPRIT, or SAGE. These algorithms, which are discussed in Chapter 5, make use of a data matrix with the array snapshots collected within the coherence time of the channel. In essence, spatial processing dynamically creates a different sector for each user and conducts a frequency/channel allocation in an ongoing manner in real time. Fig. 4.13 illustrates the beams of a fully adaptive antenna system supporting two users.

In adaptive beamforming techniques, two main strategies are distinguished. The first one is based on the assumption that part of the desired signal is already known through the

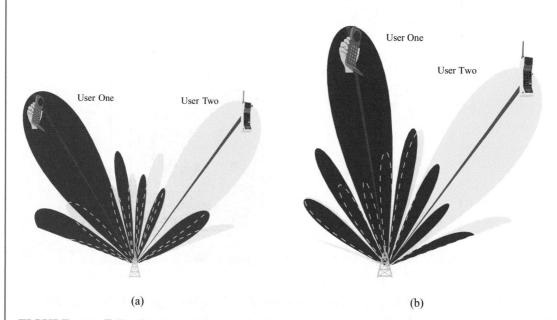

(a)   (b)

FIGURE 4.13: Fully adaptive spatial processing supporting two users on the same conventional channel simultaneously in the same cell [20].

use of a training sequence. This known signal is then compared with what is received, and the weights are then adjusted to minimize the Mean Square Error (MSE) between the known and the received signals. In this way, the beampattern can be adjusted to null the interferers. This approach optimizes the signal-to-interference ratio (SIR), and is applicable to non-line-of-sight (NLOS) environments [93]. Since the weights are updated according to the incoming signals, not only the interference is reduced but the multipath fading is also mitigated. In the second one, the directions of arrivals from all sources transmitting signals to the array antenna are first identified. The complex weights are then adjusted to produce a maximum toward the desired angle and null toward interfering signals. This strategy may turn out to be deficient in practical scenarios where there are too many DOAs due to multipaths, and the algorithms are more likely to fail in properly detecting them. This is more likely to occur in NLOS environments where there are many local scatterers close to the users and the base station, thus resulting in a wider spread of the angle of arrival [93].

Another significant advantage of the adaptive antenna systems is the ability to share spectrum. Because of the accurate tracking and robust interference rejection capabilities, multiple users can share the same conventional channel within the same cell. System capacity increases through lower inter-cell frequency reuse patterns as well as intra-cell frequency reuse. Fig. 4.13 shows how adaptive antenna approach can be used to support simultaneously two users in the same cell on the same conventional channel.

In each of the two plots, the pattern on the left is used to communicate with the user on the left while the pattern on the right is used to talk with the user on the right. The drawn lines delineate the actual direction of each signal. Notice that as the signals travel down the indicated line toward the base station, the signal from the right user arrives at a null of the left pattern or minimum gain point and vice versa. As the users move, beam patterns are constantly updated to insure these positions. The plot at the bottom of the figure shows how the beam patterns have dynamically changed to insure maximum signal quality as one user moves toward the other. Fig. 4.14 summarizes the different smart antenna concepts and the functions of each one.

4.5   SPACE DIVISION MULTIPLE ACCESS (SDMA)

A concept completely different from the previously described multiple access schemes, is the spatial division multiple access (SDMA) scheme. SDMA systems utilize techniques by which signals are distinguished at the BS based on their origin in space. They are usually used in conjunction with either FDMA, TDMA, or CDMA in order to provide the latter with the additional ability to explore the spatial properties of the signals [85]. SDMA is among the most sophisticated utilizations of smart antenna technology; its advanced spatial processing

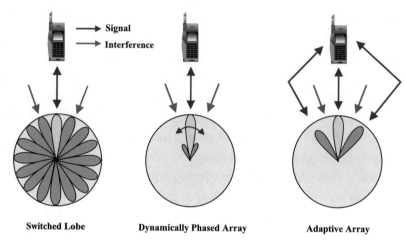

FIGURE 4.14: Different smart antenna concepts [20].

capability enables it to locate many users, creating different beams for each user, as shown in Fig. 4.15.

The SDMA scheme is based upon the concept that a signal arriving from a distant source reaches different antennas in an array at different times due to their spatial distribution [40]. This delay is utilized to differentiate one or more users in one area from those in another

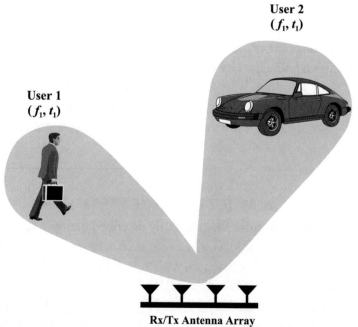

FIGURE 4.15: SDMA concept [20].

area. The scheme allows an effective transmission to take place in one cell without disturbing a simultaneous transmission in another cell. For example, conventional GSM/GPRS allows one user at a time to transmit or receive in a frequency band to the base station, where GSM/GPRS with SDMA allows multiple simultaneous transmissions in that same frequency band, multiplying the capacity of the system. CDMA system capacity is limited by its SIR, hence, with SDMA boosting the SIR in the system, more users will be allowed access by the network [94].

Filtering in the space domain can separate spectrally and temporally overlapping signals from multiple mobile units and it enables multiple users within the same radio cell to be accommodated on the same frequency and time slot [20], as illustrated in Fig. 4.15. This means that more than one user can be allocated to the same physical communication channel in the same cell simultaneously, with only separation in angle. This is accomplished by having N parallel beamformers at the base station operating independently, where each beamformer has its own adaptive beamforming algorithm to control its own set of weights and its own direction-of-arrival algorithm (DOA) to determine the time delay of each user's signal [95, 96] as shown in Fig. 4.16. Each beamformer creates a maximum toward its desired user while nulling or

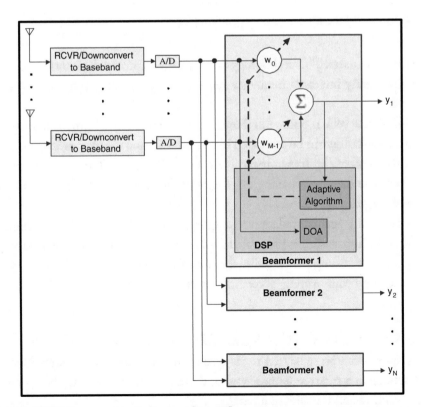

FIGURE 4.16: SDMA system block diagram [94, 95].

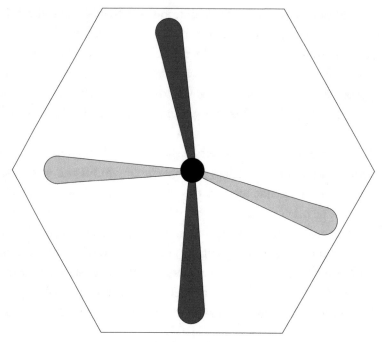

FIGURE 4.17: Channel reuse via angular separation [43].

attenuating the other users. This technology dramatically improves the interference suppression capability while greatly increases frequency reuse resulting in increased capacity and reduced infrastructure cost.

With SDMA, several mobiles can share the same frequency within a cell. Multiple signals arriving at the base station can be separated by the base station receiver as long as their angular separation is larger than the transmit/receive beamwidths [43]. This is shown in Fig. 4.17. The beams that have the same shading use the same frequency band. This technique is called *channel reuse via angular separation.*

Methods acting against fading are required for high data rate and highly reliable mobile communication systems [97]. A SDMA system is an effective measure to cope with fading, since it distinguishes radio signals in space or angular domain by using antenna directivity or beamforming according to the direction of arrival (DOA) of signals [9, 98].

4.6   ARCHITECTURE OF A SMART ANTENNA SYSTEM

Any wireless system can be separated to its reception and transmission parts. Because of the advanced functions in a smart antennas system, there is a greater need for better co-operation between its reception and transmission parts.

4.6.1   Receiver

Fig. 4.18 shows schematically the block diagram of the reception part of a wireless system employing a smart antenna with M elements. In addition to the antenna itself, it contains a radio unit, a beam forming unit, and a signal processing unit [80].

The number of elements in the array should be relatively low (the minimum required), in order to avoid unnecessarily high complexity in the signal processing unit. Array antennas can be one-, two-, and three-dimensional, depending on the dimension of space one wants to access. Fig. 4.19 shows different array geometries that can be applied in adaptive antennas implementations [80].

The first structure is used primarily for beamforming in the horizontal plane (azimuth) only. This will normally be sufficient for outdoor environments, at least in large cells. The first example (a) shows a one-dimensional linear array with uniform element spacing of Δx. Such a structure can perform beamforming in one plane within an angular sector. This is the most common structure due to its low complexity [20]. The second example (b) shows a circular array with uniform angular spacing between adjacent elements of $\Delta \varphi = 2\pi / N$, where N represents the number of elements. This structure can perform beamforming in any direction but, because of its symmetry, is more appropriate for azimuthal beamforming. The last two structures are

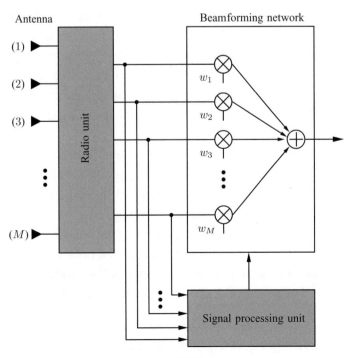

FIGURE 4.18: Reception part of a smart antenna [20].

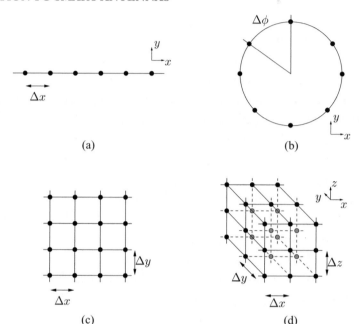

FIGURE 4.19: Different uniform array geometries for smart antennas [20].

used to perform two-dimensional beamforming, i.e. in both azimuthal and elevation angles. Such specifications are usually desirable for indoor or dense urban environments. The front view of a two-dimensional rectangular array with horizontal element spacing of Δx and vertical element spacing of Δy is shown in (c). Beamforming in the entire space, within all angles, requires some sort of cubic or spherical structure (three-dimensional configuration). The fourth example (d) shows a cubic structure with element separations of Δx, Δy, and Δy, respectively, in each direction in space.

The radio unit consists of down-conversion chains and (complex) analog-to-digital conversion (A/D). There must be M down-conversion chains, one for each of the array elements. The received signals from the mobile units are combined into one, which is the input to the remaining part of the receiver (amplifier, channel decoding, etc.).

Based on the received signal, the signal-processing unit calculates the complex weights $w_1, w_2, \ldots, w_M$ with which the received signal from each of the array elements is multiplied. These weights will determine the antenna pattern in the uplink direction. The estimate of the weights can be optimized using one of the two main criteria depending on the application and complexity:

a.   Maximization of the power of the received signal from the desired user (e.g., switched-beam or phased array), or

b.   Maximization of the SIR by suppressing the signal received from the interference sources (adaptive array).

In theory, with M antenna elements $M - 1$ sources of interference can be "nulled out", but this number will normally be lower due to the multipath propagation environment.

The method for calculating the weights differs depending on the type of optimization criterion. When the switched-beam (SB) is used, the receiver will test all the predefined weight vectors (corresponding to the beam set) and choose the best one giving the strongest received signal level. If the phased array approach (PA) is used, which consists of directing a maximum gain beam toward the strongest signal component, the weights are calculated after the direction-of-arrival (DOA) is first estimated. A number of well-documented methods exist for estimating the DOA and will be presented in Chapter 5. In the adaptive array approach (AA), where maximization of SIR is needed, the optimum weight vector (of dimension M) $\mathbf{w}_{opt}$ can be computed using a number of algorithms such as optimum combining and others that will follow.

When the beam forming is done digitally (after A/D), the beam forming and signal processing units can normally be integrated in the same unit (Digital Signal Processor, DSP). The separation in Fig. 4.18 is done to clarify the functionality. The beam forming can be performed in either at radio frequency (RF) or intermediate frequency (IF).

4.6.2   Transmitter

Normally the adaptive process is applied to the uplink/reception only (from the mobile to the base station). In that case the mobile unit consumes less transmission power, and the operational time of the battery is extended. However, the benefits of adaptation are very limited, if no beamforming is applied in the downlink transmission (from the base station to the mobile). In principle, the methods used in the uplink can be carried over the downlink [99]. The transmission part of a smart antenna system is schematically similar to its reception part as shown in Fig. 4.20.

The signal is split into N branches, which are weighted by the complex weights $w_1, w_2, \ldots, w_N$ in the lobe-forming unit. The signal-processing unit calculates suitably the weights, which form the radiation pattern in the downlink direction. The radio unit consists of D/A converters and the up-converter chains. In practice, some components, such as the antennas themselves and the DSP, will be the same as in reception. The principal difference between uplink and downlink is that since there are no smart antennas applied to the user terminals (mobile stations), there is only limited knowledge of the *Channel State Information* (CSI) available. Therefore, the optimum beamforming in downlink is difficult and the same performance as the uplink cannot be achieved.

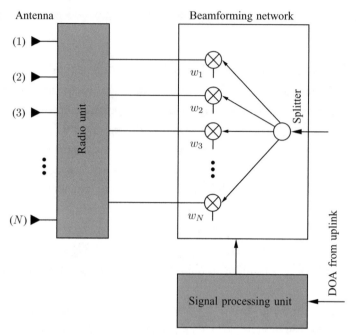

FIGURE 4.20: Transmission part of a smart antenna [20].

Typically there exist two approaches to overcome this impairment. The first one is to devise methods that do not require any CSI, but with somewhat limited performance gain. The second one is the assumption of directional reciprocity, i.e., the direction from which the signal is arrived on the uplink is closely related to the downlink CSI. This assumption has been strengthened by recent experimental results.

Physically an adaptive antenna looks very much like an ordinary antenna but has built-in electronics and control software. It cooperates with the receiver's adaptive control system in real time. It may also communicate interactively with the cellular radio network control system.

Smart antenna techniques have only recently been considered for implementation in land mobile stations and vehicle installed units because of their high system complexity and large power consumption [69, 100]. A number of smart antenna arrays for base station applications have already been proposed in [12, 13, 27]. However, only limited efforts have been yet considered for developing adaptive antenna array receivers suitable for handsets [101–103]. In fact, there exist several practical difficulties with the implementation of such a solution at the handset level [104]. These are:

i.   The space on the handset device is limited and does not allow the implementation of an antenna array with number of elements necessary enough for efficient spatial signal processing. In addition, two (or multiple) antennas in proximity may reduce the effectiveness of the antenna system due to coupling.

ii. The problem related to the movement of the mobile that provides an omnidirectional scenario.

iii. The cost and the complexity of the implementation at every mobile is much greater than the implementation at each base radio station.

Besides these difficulties, the adaptive algorithm for signal processing at the handset must be fast; however it needs only a few simple calculations, and requires a simple hardware implementation [104]. To justify further research efforts in employing multiple antennas at handsets, the gain in performance should be large enough to offset the additional cost and power consumption [69]. Finally, it can be stressed that the use of digital beamforming antennas, both in satellites and in land-fixed and mobile units, remains a challenge for future satellite communication systems.

4.7 BENEFITS AND DRAWBACKS

The introduction of smart antennas is expected to have a large impact on the performance of cellular communications networks. It will also affect many aspects of both the planning and deployment of mobile systems. The great interest in smart antennas is the increase in capacity and range. In densely populated areas the main source of noise is the interference from other users. The deployment of adaptive arrays is to simultaneously increase the useful received signal level and lower the interference level, thus providing significant improvement in the Signal to Interference Ratio (SIR). An immediate impact to the increase of the SIR is the possibility for reduced frequency reuse distance. This will lead to a large capacity increase since more carriers can be allocated per cell. An immediate advantage will be noticed in TDMA systems (GSM) which are more concerned about increased SIR. An example is shown in Fig. 4.21, where the traditional seven-cell cluster has been reduced to a three-cell cluster. This will lead to a capacity increase of 7/3.

Using smart antennas, an increase of the range of coverage by a base station is possible since they are able to focus their energy toward the intended users instead of directing and wasting it in other unnecessary directions. In other words, smart antennas are more directive than traditional sectorized or omnidirectional antennas. Thus, base stations can be placed further apart, potentially leading to more cost-efficient deployment [20]. Therefore, in rural and sparsely populated areas, where radio coverage rather than capacity is more important, smart antenna systems are also well-suited [13]. Moreover, using transmit and receive beams that are directed toward the mobile user of interest, the multipath [105] and the

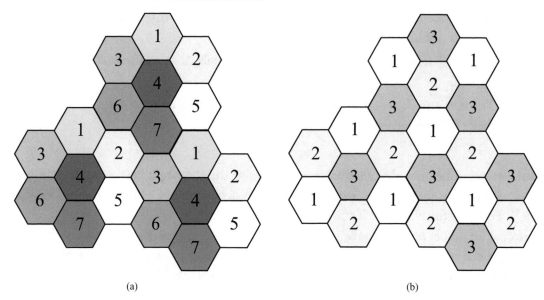

(a)   (b)

FIGURE 4.21: (a) Traditional 7-cell cluster and (b) possible 3-cell cluster enabled by interference reduction like when employing smart antennas [20].

inter-symbol-interference, due to multipath propagation present in mobile radio environments, are mitigated.

Another added advantage of smart antenna systems is security. In a society that becomes more dependent on conducting business and distributing personal information, security is an important issue. Smart antennas make it more difficult to tap a connection because the intruder must be positioned in the same direction as the user as "seen" from the base station to successfully tap a connection [13].

Finally, due to the spatial detection nature of smart antenna systems, the network will have access to spatial information about users [20]. This information may be exploited in estimating the positions of the users much more accurately than in existing networks. Consequently, exact positioning can be used in services to locate humans in case of emergency calls or for any other location-specific service [20].

Although the benefits of using smart antennas are considered many, there also exist some important drawbacks. A smart antenna transceiver is much more complicated than a traditional base station transceiver [80]. Separate transceiver chains are needed for each of the array antenna elements and accurate real-time calibration of each of them is required. Moreover, adaptive beamforming is a computationally intensive process; thus the smart antenna base station must include very powerful numeric processors and control systems.

FIGURE 4.22: Picture of an eight-element array antenna at 1.8 GHz. (Antenna property of Telia Research AB, Sweden) [80].

Smart antenna base stations will no doubt be much more expensive than conventional base stations.

Even though smart antennas are mainly a radio technology, it will unavoidably put new demands on network functions such as resource and mobility management. SDMA involves different users using the same physical communication channel in the same cell, separated only by angle. When angular collisions between these users occur, one of them must quickly switch to another channel so that the connection is not broken.

For the smart antenna to obtain a reasonable gain, an antenna array with several elements is necessary. Typically arrays consisting of six to ten horizontally separated elements have been suggested for outdoor mobile environments. The necessary element spacing is 0.4–0.5 wavelengths. An eight-element antenna, for example, would be approximately 1.2 meters wide at a frequency of 900 MHz and 60 cm at 2 GHz. With a growing public demand for less visible base stations, geometries with size of several wavelengths (corresponding to current carrier frequencies used), although not excessive, could provide a problem. Fig. 4.22, showing a picture of an eight-element antenna array operating at 1.8 GHz, reinforces the argument.

4.8   BASIC PRINCIPLES

The diagram of Fig. 4.23 shows the principal system elements of a "smart" antenna system.

The smart antenna consists of the *sensor array*, the *patternforming network*, and the *adaptive processor*:

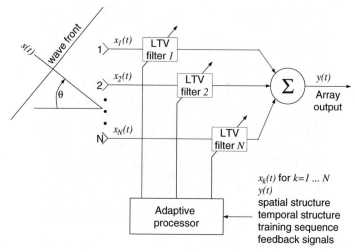

FIGURE 4.23: Functional diagram of an N element smart antenna [43].

i. *Sensor Array*: The sensor array consists of N sensors designed to receive (and transmit) signals. The physical arrangement of the array (linear, circular, etc.) can be chosen arbitrarily, depending on the required specifications. However, it places fundamental limitations on the capability of the smart antenna.

ii. *Patternforming Network*: The output of each of the N sensor elements is fed into the patternforming network, where the outputs are processed by linear time-variant (LTV) filters. These filters determine the directional pattern[1] of the smart antenna. The outputs of the LTV filters are then summed to form the overall output $y(t)$. The complex weights of the LTV filters are determined by the adaptive processor.

iii. *Adaptive Processor*: The adaptive processor determines the complex weights of the patternforming network. The signals and known system properties used to compute the weights include the following:
- The signals received by the sensor array, i.e., $x_n(t)$, $n = 1, 2, \ldots, N$.
- The output of the smart antenna, i.e., $y(t)$.
- The *spatial structure* of the sensor array.
- The *temporal structure* of the received signal.
- Feedback signals from the mobiles.
- Network topology.

[1]The relative sensitivity of response to signals for a specified frequency from various directions.

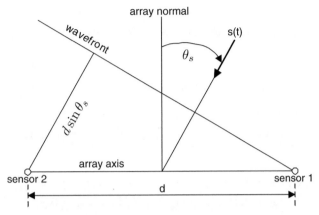

FIGURE 4.24: A uniform linear array (ULA) with two sensor elements along with an impinging uniform plane EM wave [43].

We can explain the functioning principle of a "smart" antenna using a simple example. In the example, we consider a *uniform linear array* (ULA) consisting of two identical omnidirectional sensors as shown in Fig. 4.24. We assume that a signal $s(t)$ is generated by a source in the far-field of the "smart" antenna. The impinging signal on the *sensor array* is approximately a uniform plane wave, as shown in Fig. 4.24. With respect to sensor 1, sensor 2 experiences a time delay of

$$\Delta \tau = \frac{d \sin \theta_s}{v_0} \tag{4.1}$$

where d is the spacing between the two elements and v_0 the wave speed. Similarly, knowing d and measuring $\Delta \tau$, the angle θ_s of the direction of arrival is found using

$$\theta_s = \sin^{-1} \left[\frac{v_0 \Delta \tau}{d} \right]. \tag{4.2}$$

If $s(t)$ is a narrowband signal with carrier frequency f_c, then the time delay $\Delta \tau$ corresponds to a phase shift of

$$\Delta \psi = 2\pi \frac{d \sin \theta_s}{\lambda_c} \tag{4.3}$$

where λ_c is the wavelength corresponding to the carrier frequency, i.e., $\lambda_c = \frac{v_0}{f_c}$. Clearly, for an incoming signal from a direction perpendicular to the array normal ($\theta_s = 0$), both the time delay and phase shift between the two sensors are zero.

Now, let us assume that an interfering signal $n(t)$ with the same carrier frequency f_c impinges on the array. As an example, the directions of $s(t)$ and $n(t)$ are set to $0°$ and $30°$, respectively. We denote the complex sensor weights as $w_1 = w_{1,1} + jw_{1,2}$ and $w_2 = w_{2,1} + jw_{2,2}$, respectively. Thus, the array output due to $s(t)$ is given by

$$\begin{aligned} S(t) &= s(t)\left[(w_{1,1} + jw_{1,2}) + (w_{2,1} + jw_{2,2})\right] \\ &= s(t)\left[(w_{1,1} + w_{2,1}) + j(w_{1,2} + w_{2,2})\right]. \end{aligned} \tag{4.4}$$

According to (4.3), for interelement spacing $d = \frac{\lambda}{2}$ and $\theta_s = 30°$, sensor 2 exhibits a phase lag of $\Delta\psi = \frac{\pi}{2}$ with respect to sensor 1. Thus, the array output due to $n(t)$ is given by

$$\begin{aligned} N(t) &= n(t)\left[(w_{1,1} + jw_{1,2}) + e^{-j\frac{\pi}{2}}(w_{2,1} + jw_{2,2})\right] \\ &= n(t)\left[(w_{1,1} + w_{2,2}) + j(w_{1,2} - w_{2,1})\right]. \end{aligned} \tag{4.5}$$

The goal of the "smart" antenna is to cancel out completely the interfering signal $n(t)$ and fully recover the desireed signal $s(t)$. To achieve this objective, using (4.4) and (4.5), it is necessary that

$$w_{1,1} + w_{2,1} = 1 \tag{4.6a}$$
$$w_{1,2} + w_{2,2} = 0 \tag{4.6b}$$
$$w_{1,1} + w_{2,2} = 0 \tag{4.6c}$$
$$w_{1,2} - w_{2,1} = 0. \tag{4.6d}$$

Solution of (4.6) yields

$$w_{1,1} = w_{1,2} = w_{2,1} = \frac{1}{2} \quad\text{and}\quad w_{2,2} = -\frac{1}{2} \tag{4.7}$$

or

$$w_1 = \frac{1}{2}(1 + j) \tag{4.8a}$$

$$w_2 = \frac{1}{2}(1 - j). \tag{4.8b}$$

For every array antenna, its *steering vector* can be defined. The steering vector contains the responses of all elements of the array to a source with a single frequency component of unit power [40]. Since the array response is different in different directions, a steering vector is associated with each directional source. The array geometry defines the uniqueness of this association [106]. For an array of identical elements, each component of this vector has unit magnitude. The phase of its nth component is equal to the phase difference between signals induced on the nth element and the reference element due to the source associated with the steering vector [40]. The reference element usually is set to have zero phase. This vector is also known as the *space vector* since each component of this vector denotes the phase delay caused by

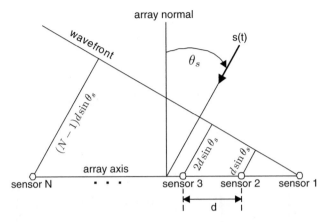

FIGURE 4.25: A ULA of N elements and element spacing d along with an impinging planar wavefront.

the spatial position of the corresponding element of the array. It is also referred to as the *array response vector*, as it measures the response of the array due to the source under consideration [40].

For example, the steering vector of an N-element ULA with spacing d between adjacent elements, as shown in Fig. 4.25, is given by

$$\mathbf{a}(\theta) = \left[1, e^{-j\frac{2\pi d}{\lambda}\sin\theta_s}, \ldots, e^{-j(N-1)\frac{2\pi d}{\lambda}\sin\theta_s}\right]^T. \tag{4.9}$$

Once the steering vector for an array antenna is derived, its radiation pattern is formed by applying to each entry of the steering vector the excitation, in amplitude and phase, of the corresponding antenna element.

The normalized *radiation pattern* of an N-element ULA of omnidirectional elements with inter-element spacing d is given in dB by

$$G(\theta) = 10\log_{10}\left\{\frac{|\mathbf{w}^T\mathbf{a}(\theta)|^2}{\mathbf{w}^H\mathbf{w}}\right\}$$

$$= 10\log_{10}\left\{\frac{\left|\sum_{n=0}^{N-1} w_n\exp\left(-\frac{j2\pi nd\sin\theta}{\lambda_c}\right)\right|^2}{\mathbf{w}^H\mathbf{w}}\right\} \tag{4.10}$$

where $\mathbf{w}$ is the N-dimensional vector consisting of the complex weights of the antenna elements. Fig. 4.26 shows the normalized radiation patter for a two-element antenna array without any weighting in the pattern forming network.

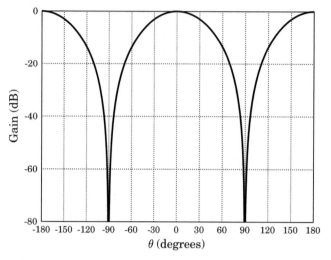

FIGURE 4.26: Normalized radiation pattern for a non-weighted two-sensor array.

Fig. 4.27 shows the normalized radiation pattern for a two-element antenna array when the weights of (4.7) are applied in the pattern forming network. It is seen that now a null is placed exactly at an azimuth of $30°$, the direction of the interferer.

By this simple example, we see how the complex weights of an array of antenna elements can be adjusted such that to completely null the interfering signal $n(t)$ and the output to be equal to the desired signal $s(t)$. This model can be easily expanded in the case of a ULA with N elements. Such an array, can recover the desired signal $s(t)$ and fully cancel out $N-1$

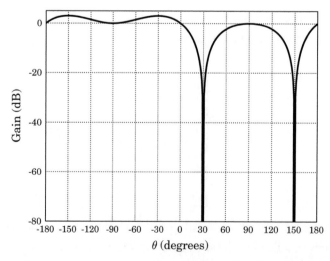

FIGURE 4.27: Normalized radiation pattern for a weighted two-sensor array.

interfering signals. To demonstrate this capability, let us assume a scenario in which the desired source and $N-1$ interfering sources transmit signals towards an N-element ULA. The useful signal $s(t)$ arrives from angle θ_s. The phase shift between the mth sensor and the first sensor due to $s(t)$ is given by

$$\Delta \psi_{s,m} = 2\pi \frac{(m-1)d \sin \theta_s}{\lambda_c}, \quad m = 1, 2, \ldots, N. \tag{4.11}$$

The interfering signals arrive from angles $\theta_1, \theta_2, \ldots, \theta_{N-1}$. The phase shift between the mth sensor and the first sensor due to the nth interfering signal is given by

$$\Delta \psi_{n,m} = 2\pi \frac{(m-1)d \sin \theta_n}{\lambda_c}, \quad \begin{cases} m = 1, 2, \ldots, N \\ n = 1, 2, \ldots, N-1. \end{cases} \tag{4.12}$$

The complex sensor weights are $w_1 = w_{1,1} + jw_{1,2}, w_2 = w_{2,1} + jw_{2,2}, \ldots, w_N = w_{N,1} + jw_{N,2}$. Therefore, the array output due to $s(t)$ is given by

$$\begin{aligned} S(t) = s(t) \big[(w_{1,1} + jw_{1,2}) + e^{-j\Delta\psi_{s,2}} (w_{2,1} + jw_{2,2}) + \ldots + \\ + e^{-j\Delta\psi_{s,N}} (w_{N,1} + jw_{N,2}) \big] \end{aligned} \tag{4.13}$$

and due to the nth interfering signal $n_n(t)$ is given by

$$\begin{aligned} N_n(t) = n_n(t) \big[(w_{1,1} + jw_{1,2}) + e^{-j\Delta\psi_{n,2}} (w_{2,1} + jw_{2,2}) + \ldots + \\ + e^{-j\Delta\psi_{n,N}} (w_{N,1} + jw_{N,2}) \big], \quad n = 1, 2, \ldots, N-1. \end{aligned} \tag{4.14}$$

The total array output is given by

$$y(t) = S(t) + N(t) = S(t) + \sum_{n=1}^{N-1} N_n(t). \tag{4.15}$$

There are $2N$ unknowns to be determined, the N real parts and the N imaginary parts of the N complex weights such that the N-element ULA to recover the desired signal and fully cancel out the $N-1$ interfering signals. Equivalently, the total array output must be equal to the useful signal, or

$$y(t) = s(t). \tag{4.16}$$

For convenience, we define the $2N$-dimensional vector $\mathbf{w}$ as

$$\mathbf{w} = \left[\mathbf{w}_1^T, \mathbf{w}_2^T \right]^T \tag{4.17}$$

where both $\mathbf{w}_1$ and $\mathbf{w}_2$ are N-dimensional vectors consisting of the real and imaginary parts of the complex weights of the N array elements, respectively, or

$$\mathbf{w}_1 = [w_{1,1}, w_{2,1}, \ldots, w_{N,1}]^T \qquad (4.18)$$
$$\mathbf{w}_2 = [w_{1,2}, w_{2,2}, \ldots, w_{N,2}]^T. \qquad (4.19)$$

Furthermore, we define the four matrices $\mathbf{R}_1, \mathbf{R}_2, \mathbf{R}_3,$ and $\mathbf{R}_4$, all $\in \mathbb{R}^{N \times N}$, as

$$\mathbf{R}_1 = \begin{bmatrix} 1 & \cos(\Delta\psi_{s,2}) & \ldots & \cos(\Delta\psi_{s,N}) \\ 1 & \cos(\Delta\psi_{1,2}) & \ldots & \cos(\Delta\psi_{1,N}) \\ \vdots & \vdots & \ddots & \vdots \\ 1 & \cos(\Delta\psi_{N-1,2}) & \ldots & \cos(\Delta\psi_{N-1,N}) \end{bmatrix} \qquad (4.20)$$

$$\mathbf{R}_2 = \begin{bmatrix} 0 & \sin(\Delta\psi_{s,2}) & \ldots & \sin(\Delta\psi_{s,N}) \\ 0 & \sin(\Delta\psi_{1,2}) & \ldots & \sin(\Delta\psi_{1,N}) \\ \vdots & \vdots & \ddots & \vdots \\ 0 & \sin(\Delta\psi_{N-1,2}) & \ldots & \sin(\Delta\psi_{N-1,N}) \end{bmatrix} \qquad (4.21)$$

$$\mathbf{R}_3 = -\mathbf{R}_2 \qquad (4.22)$$

$$\mathbf{R}_4 = \mathbf{R}_1. \qquad (4.23)$$

Using (4.13), (4.14), (4.15), (4.17), (4.19), and (4.23), the total array output is obtained by

$$y(t) = [s(t), n_1(t), n_2(t), \ldots, n_{N-1}(t)] \{[\mathbf{R}_1 \ \mathbf{R}_2] + j [\mathbf{R}_3 \ \mathbf{R}_4]\} \mathbf{w}. \qquad (4.24)$$

From (4.24), solution to (4.16) is given by

$$\mathbf{w} = \mathbf{R}^{-1}\mathbf{M}. \qquad (4.25)$$

where the matrix $\mathbf{R} \in \mathbb{R}^{2N \times 2N}$ is given by

$$\mathbf{R} = \begin{bmatrix} \mathbf{R}_1 & \mathbf{R}_2 \\ \mathbf{R}_3 & \mathbf{R}_4 \end{bmatrix}. \qquad (4.26)$$

The $2N$-dimensional vector $\mathbf{M}$ can be written as

$$\mathbf{M} = \begin{bmatrix} \mathbf{M}_1^T, \mathbf{M}_2^T \end{bmatrix}^T \qquad (4.27)$$

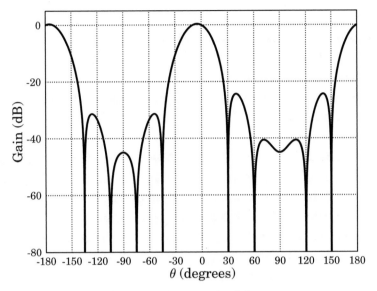

FIGURE 4.28: Normalized radiation pattern for a weighted five-sensor array.

where both $\mathbf{M}_1$ and $\mathbf{M}_2$ are N-dimensional vectors. $\mathbf{M}_1$ is written as

$$\mathbf{M}_1 = [1, 0, 0, \ldots, 0]^T \tag{4.28}$$

and is interpreted as preserving the real part of the useful signal $s(t)$ and cancelling out the real parts of all the interfering signals $n_n(t)$, $n = 1, 2, \ldots, N - 1$. Similarly, $\mathbf{M}_2$ is written as

$$\mathbf{M}_2 = [0, 0, \ldots, 0]^T. \tag{4.29}$$

and is interpreted as nulling out the imaginary parts of the useful signal $s(t)$ and all the interfering signals $n_n(t)$, $n = 1, 2, \ldots, N - 1$.

Fig. 4.28 represents a more complicated example of a patternforming network. In this example, we assume a ULA with five omnidirectional sensors. The incoming signal of interest arrives from angle $\theta_s = 0°$ and the four interfering signals from angles $\theta_1 = -75°$, $\theta_2 = -45°$, $\theta_3 = 30°$, and $\theta_4 = 60°$.

In Fig. 4.26, Fig. 4.27, and Fig. 4.28, we see that the radiation pattern is identical in directions symmetric about the linear array axis, or

$$G(\theta) = G(\pi - \theta), \quad 0 \le \theta \le \pi, \text{ and} \tag{4.30a}$$
$$G(\theta) = G(-\pi + \theta), \quad -\pi \le \theta \le 0. \tag{4.30b}$$

For a plane wave arriving from angle either θ or $180° - \theta$, each element in the array experiences identical time delay. This observation is useful for the analysis that follows.

At this point, one may wonder: Is the solution to the complex elements weights always feasible? The answer to this question is yes, if and only if the set of $2N$ equations which yield the solution to $\mathbf{w}$ are independent, or equivalently the matrix $\mathbf{R}$ in (4.25) is of full rank $2N^2$. For this to happen, there should not exist any pair of angles of arrivals symmetrical about the axis of the linear array. Any pair of angles of arrivals symmetrical about the axis of the linear array reduces the rank of the matrix $\mathbf{R}$ by 2 (since both the real and imaginary parts of the complex weights are taken into account).

In case the number of interferers is less than $N - 1$, one may impose an additional number of required nulls, which will be of no practical interest, such that none possible combination of pairs of angle of arrivals is symmetric about the axis of the array.

Similarly, in case the number of interferers is greater than $N - 1$, there should be pairs of angle of arrivals symmetric about the axis of the array. The number of this pairs must be exactly the excess of the total number of interfering signals from $N - 1$. A smaller number leads to an array design that cannot cancel out all the present interferers.

This flexibility of an N-element array to be able to fix the pattern at $N - 1$ places is known as the degree of freedom of the array [40]. For a ULA, this is similar to an $N - 1$ degree polynomial of $N - 1$ adjustable coefficients, with the first coefficient having the value of unity.

4.9   MUTUAL COUPLING EFFECTS

In the previous example, we ignored any array imperfections and the radiation influence between each other elements in the array. However, when the radiating elements in the array are in the vicinity to each other, the radiation characteristics, such as the input impedance and the radiation pattern, of an excited antenna element are influenced by the presence of the others. This effect is known as *mutual coupling*, and it can have a deleterious impact on the performance of a smart antenna array. Mutual coupling usually causes the maximum and nulls of the radiation pattern to shift; consequently, the DOA algorithm and the beamforming algorithm produce inaccurate results unless this effect is taken into account. Furthermore, this detrimental effect intensifies as the interelement spacing is reduced [58, 59, 108]. For more details on the effects of mutual coupling on the performance of adaptive arrays, and compensation techniques, the interested reader is referred to the literature [108–113]. However, a simple example follows to illustrate the effects of mutual coupling on adaptive beamforming. More on mutual coupling later in Section 7.3 *Mutual coupling* of Chapter 7: Integration and Simulation of Smart Antennas.

---

[2]The rank of a matrix is the minimum number of linearly independent rows or columns in the matrix, whichever is less [107].

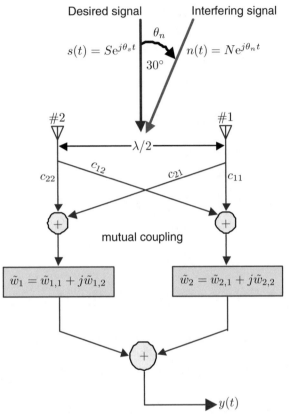

Desired signal   Interfering signal

θ_n

$s(t) = Se^{j\theta_s t}$ $n(t) = Ne^{j\theta_n t}$

$30°$

#2   #1

$\lambda/2$

c_{22} c_{12} c_{21} c_{11}

mutual coupling

$\tilde{w}_1 = \tilde{w}_{1,1} + j\tilde{w}_{1,2}$ $\tilde{w}_2 = \tilde{w}_{2,1} + j\tilde{w}_{2,2}$

$y(t)$

FIGURE 4.29: Two-element ULA receiving a desired signal (SOI) at $\theta_s = 0°$ and an interferer (SNOI) at $\theta_n = 30°$ in the presence of mutual coupling.

Let us consider again the example of the uniform linear array with two sensor elements of Fig. 4.24. The spacing between the elements is half-wavelength ($d = \lambda/2$) and the desired signal (SOI) is arriving at an angle $\theta_s = 0°$. Also, it is required to tune out an interferer (SNOI) at $\theta_n = 30°$. For simplicity, the elements in the array are assumed to be omnidirectional and the impinging signals sinusoids. However, this time mutual coupling between the radiating elements is considered. Fig. 4.29 illustrates this configuration [59].

The output $y(t)$ of the array due to the desired signal $s(t)$ is analyzed first, followed by the output due to the interferer $n(t)$. Assuming an angular frequency ω_0, the output $y(t)$ of the array due to $s(t)$ is [59]

$$y(t) = Se^{j\omega_0 t} [(c_{11} + c_{12}) \tilde{w}_1 + (c_{21} + c_{22}) \tilde{w}_2] \tag{4.31}$$

where c_{11}, c_{12}, c_{21}, and c_{22} represent, respectively, the mutual coupling coefficients. These coefficients describe the way an element is affected due to the presence of another. Therefore, for the output $y(t)$ to be equal only to the desired signal, $s(t)$, it is necessary that

$$(c_{11} + c_{12})\, \tilde{w}_1 + (c_{21} + c_{22})\, \tilde{w}_2 = 1. \tag{4.32}$$

On the other hand, the output $y(t)$ due to the interfering signal $n(t)$ is given as [59]

$$y(t) = N\left[e^{(j\omega_0 t + \pi/4)}\left(c_{11}\tilde{w}_1 + c_{21}\tilde{w}_2\right) + e^{(j\omega_0 t - \pi/4)}\left(c_{12}\tilde{w}_1 + c_{22}\tilde{w}_2\right)\right] \tag{4.33}$$

where the $+\pi/4$ and $-\pi/4$ terms appear due to the phase lead and lag of elements 1 and 2, respectively, in reference to the impinging interfering signal on the array midpoint. Because

$$e^{(j\omega_0 t \pm \pi/4)} = \frac{e^{j\omega_0 t}}{\sqrt{2}}\left(1 \pm j\right) \tag{4.34}$$

(4.33) can be written as

$$y(t) = N e^{j\omega_0 t} \frac{\sqrt{2}}{2}\left\{\left[(1 + j)c_{11} + (1 - j)c_{12}\right]\tilde{w}_1 + \left[(1 + j)c_{21} + (1 - j)c_{22}\right]\tilde{w}_2\right\}. \tag{4.35}$$

Therefore, for the output response to reject totally the interference, it is necessary that

$$\left[(1 + j)c_{11} + (1 - j)c_{12}\right]\tilde{w}_1 + \left[(1 + j)c_{21} + (1 - j)c_{22}\right]\tilde{w}_2 = 0. \tag{4.36}$$

Solving simultaneously the linear system of complex equations for $\tilde{w}_1$ and $\tilde{w}_2$ in (4.32) and (4.36) yields

$$\tilde{w}_1 = \frac{1}{2}\left(\frac{c_{22} - c_{21}}{c_{11}c_{22} - c_{12}c_{21}} + j\frac{c_{22} + c_{21}}{c_{11}c_{22} - c_{12}c_{21}}\right) \tag{4.37a}$$

$$\tilde{w}_2 = \frac{1}{2}\left(\frac{c_{11} - c_{12}}{c_{11}c_{22} - c_{12}c_{21}} - j\frac{c_{11} + c_{12}}{c_{11}c_{22} - c_{12}c_{21}}\right). \tag{4.37b}$$

Note that the above-computed weights in the presence of mutual coupling are related to those weights in the absence of mutual coupling by [59]

$$\tilde{w}_1 = w_1\left(\frac{c_{22}}{c_{11}c_{22} - c_{12}c_{21}} + j\frac{c_{21}}{c_{11}c_{22} - c_{12}c_{21}}\right) \tag{4.38a}$$

$$\tilde{w}_2 = w_2\left(\frac{c_{11}}{c_{11}c_{22} - c_{12}c_{21}} - j\frac{c_{12}}{c_{11}c_{22} - c_{12}c_{21}}\right) \tag{4.38b}$$

where w_1 and w_2 are the computed weights in the absence of mutual coupling as derived in (17).

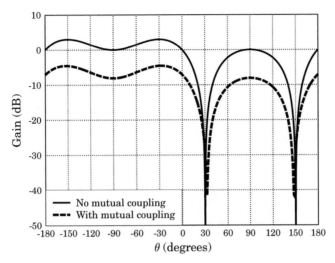

FIGURE 4.30: Comparison of array factors in the absence and presence of mutual coupling.

Based on the formulation in [108], let us now assume that the values for c_{11}, c_{12}, c_{21}, and c_{22} are given by [17]

$$c_{11} = c_{22} = 2.37 + j0.34 \qquad (4.39a)$$
$$c_{12} = c_{21} = -0.13 - j0.0517. \qquad (4.39b)$$

Then, using the weights of (17) and the mutual coupling coefficients of (41), the computed altered weights in the presence of mutual coupling are

$$\tilde{w}_1 = \tilde{w}_{1,1} + j\tilde{w}_{1,2} = 0.250 + j0.167 \qquad (4.40a)$$
$$\tilde{w}_2 = \tilde{w}_{2,1} + j\tilde{w}_{2,2} = 0.189 - j0.224. \qquad (4.40b)$$

Based on the weights of (42) and those of (17), the computed patterns with and without mutual coupling are displayed in Fig. 4.30.

It is apparent that mutual coupling plays a significant role in the formation of the pattern. For example, in the presence of coupling, the pattern minimum (in the direction of the SNOI) is displaced approximately at $\theta = 32.4°$ and at a level of approximately 41.57 dB below the value of the pattern in the direction of the SOI, while in the absence of coupling, the null is exactly at $\theta = 30°$ and at a level of nearly $-\infty$ dB.

CHAPTER 5

DOA Estimation Fundamentals

In many practical signal processing problems, the objective is to estimate from a collection of noise "contaminated" measurements a set of *constant* parameters upon which the underlying true signals depend [21]. Moreover, as clearly understood from the previous chapter, the accurate estimation of the *direction of arrival* of all signals transmitted to the adaptive array antenna contributes to the maximization of its performance with respect to recovering the signal of interest and suppressing any present interfering signals. The same problem of determining the DOAs of impinging wavefronts, given the set of signals received at an antenna array from multiple emitters, arises also in a number of radar, sonar, electronic surveillance, and seismic exploration applications.

The resolution properties of antenna arrays have been extensively investigated by many researchers. A significant portion of these efforts has been devoted to the estimation of performance bounds for any given array geometry. The reason is the comparison of the performance of the DOA estimation and beamforming methods to several basic array geometries. The theoretical performance bound studies are concerned mostly with the derivation of the *Cramér–Rao lower bound* (CRLB) for DOA estimation variance given an arbitrary array geometry. The CRLB gives the variance lower bound of the unbiased estimator of a parameter or parameter vector [110]. In [114], there are detailed discussions and derivations, as well, of the CRLB for various scenarios.

In the case of the DOA estimation, the CRLB provides the metric to compare the arrays in an algorithm-independent way, because specific algorithms may exploit special properties of certain geometries and thus, performance comparisons using any given algorithm cannot be considered conclusive. In the studies by Messer et al. [115] and Mirkin and Sibul [116], as well, CRLB expressions for azimuth and elevation angles estimates of a single source using arbitrary two-dimensional array geometries are derived. Nielsen [117] and Goldberg and Messer [118], as well, have derived single source DOA estimation and CRLB expressions are derived for arbitrary three-dimensional antenna array geometries while in Dogandzic and Nehorai [119], CRLB expressions are derived for the range, velocity and DOA estimates of a single signal source when arbitrary 3D antenna array geometries are used. It is also shown that the CRLBs depend only on the "moment of inertia" of the array geometry. Furthermore, Ballance and Shaffer [120], and Bhuyan and Schultheiss [121], have provided CRLB expressions when there are two signal sources in the system. To the best of our knowledge, no result for CRLB expressions for systems with three or more signals or more sources can be found in the published literature so far.

5.1   INTRODUCTION

In this chapter, we discuss the DOA estimation algorithms which are directly associated with the received signals. Data from an array of sensors are collected, and the objective is to locate point sources assumed to be radiating energy that is detectable by the sensors. Mathematically, such problems are modeled using *Green's functions* for the particular differential operator that describes the physics of radiation propagation from the sources to the sensors [122]. Although most of the so-called *high resolution direction finding* (DF) algorithms (e.g., MUSIC [123], maximum likelihood, autoregressive modelling techniques, etc.) have been presented in the context of estimating a single angle per emitter (e.g., azimuth only), generalizations to the azimuth/elevation case are relatively straightforward. Additional parameters, such as frequency, polarization angle, and range can also be incorporated, provided that the response of the array is known as a function of these parameters. A simple example of such an application, for the DOA to be the parameter for estimation, is depicted in Fig. 5.1, where signals from two sources impinge on an array of three coplanar receivers. The patterns associated with each receiver indicate their relative directional sensitivity. For the intended application, a few *reasonable* assumptions can be invoked to make the problem analytically tractable. The transmission medium is assumed to be isotropic and nondispersive and the sources are located in the *far-field* of the array so that the radiation impinging on the array is in the form of sum of *plane waves* [122]. Otherwise, for closely located sources (in the *near-field* of the array) the wavefronts would possess the analogous curvature.

The main difficulties associated with these methods are that both computational and storage costs tend to increase rapidly with the dimension of the parameter vector. The increased costs are usually prohibitive even for the two-dimensional (2D) case, and the result is that, in practice, systems typically employ *nonparametric* techniques (e.g., beamforming) to solve what in reality are *parametric* problems. Though these classical DF techniques are less complicated, their performance is known to be poor [124].

In general, the DOA estimation algorithms can be categorized into two groups; the *conventional* algorithms and the *subspace* algorithms. Before we proceed in presenting them, we first need to introduce the concepts of the *array response vector* and the *signal autocovariance matrix*.

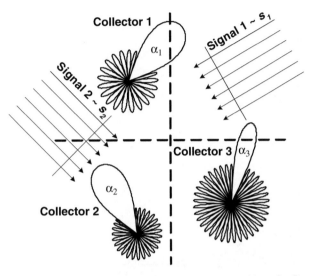

FIGURE 5.1: Illustration of a simple source location estimation problem [21].

5.2   THE ARRAY RESPONSE VECTOR

Assuming that an antenna array is composed of identical isotropic elements, each element receives a time-delayed version of the same plane wave with wavelength λ. In other words, each element receives a phase-shifted version of the signal. For example, with a *uniform linear array* (ULA), as shown in Fig. 5.2, the relative phases are also uniformly spaced, with $\psi = \frac{2\pi}{\lambda} d \sin\theta$ being the relative phase difference between adjacent elements.

The vector of relative phases is referred to as the *steering vector* (SV), also mentioned in the previous chapter. A more general concept is the *array response vector* (ARV) which is the response of an array to an incident plane wave. It is a combination of the steering vector and the response of each individual element to the incident wave. The general normalized ARV expression for a three-dimensional array of N elements is

$$\mathbf{a}(\theta, \phi) = \begin{bmatrix} G_1(\theta, \phi)e^{-j\boldsymbol{\beta}\cdot\mathbf{r}_1} \\ G_2(\theta, \phi)e^{-j\boldsymbol{\beta}\cdot\mathbf{r}_2} \\ \vdots \\ G_N(\theta, \phi)e^{-j\boldsymbol{\beta}\cdot\mathbf{r}_N} \end{bmatrix} \tag{5.1}$$

where $\boldsymbol{\beta}$ is the vector wavenumber of the incident plane wave ($\boldsymbol{\beta} = [\sin\theta\cos\phi, \sin\theta\sin\phi, \cos\theta]$ in cartesian coordinates), $\mathbf{r}_i = [x_i, y_i, z_i]$ is the three-dimensional position vector of the ith element in the array and $G_i(\theta, \phi)$ is the gain of the ith element toward the direction (θ, φ), where θ and φ are the elevation and azimuth angles, respectively. For an array

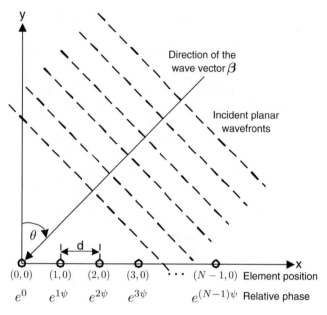

FIGURE 5.2: Array response vector for a uniform linear array [19].

of isotropic radiators, the ARV simplifies to the SV:

$$\mathbf{a}(\theta, \phi) = \left[e^{-j\boldsymbol{\beta}\cdot\mathbf{r}_1}, e^{-j\boldsymbol{\beta}\cdot\mathbf{r}_2}, \ldots, e^{-j\boldsymbol{\beta}\cdot\mathbf{r}_N} \right]^T. \tag{5.2}$$

In the paper by Chambers *et al.* [125], the CRLB for the azimuth and elevation DOA estimation variances for an arbitrary three-dimensional array are given by:

$$\mathrm{CRLB}(\theta) = \frac{1 + ASNR}{2N(ASNR)^2} \frac{AV_{\phi\phi}}{AV_{\theta\theta} AV_{\phi\phi} - AV_{\theta\phi}^2} \tag{5.3a}$$

$$\mathrm{CRLB}(\phi) = \frac{1 + ASNR}{2N(ASNR)^2} \frac{AV_{\theta\theta}}{AV_{\theta\theta} AV_{\phi\phi} - AV_{\phi\theta}^2} \tag{5.3b}$$

where ASNR is the *antenna signal-to-noise ratio* and

$$AV_{\theta\theta} = \frac{\partial \mathbf{a}^H}{\partial \theta} \frac{\partial \mathbf{a}}{\partial \theta}, \tag{5.4a}$$

$$AV_{\phi\phi} = \frac{\partial \mathbf{a}^H}{\partial \phi} \frac{\partial \mathbf{a}}{\partial \phi}, \text{ and} \tag{5.4b}$$

$$AV_{\theta\phi} = AV_{\phi\phi} = \frac{\partial \mathbf{a}^H}{\partial \theta} \frac{\partial \mathbf{a}}{\partial \phi} = \frac{\partial \mathbf{a}^H}{\partial \phi} \frac{\partial \mathbf{a}}{\partial \theta}. \tag{5.4c}$$

5.3   RECEIVED SIGNAL MODEL

Let us first assume that K uncorrelated sources transmit signals to an N-element antenna array. It is assumed here that the array response for each signal is a function of only one angle parameter (θ). For our analysis we will employ the well-established *narrowband* data model. The model inherently assumes that as the signal wavefronts propagate across the array, the *envelop* of the signal is essentially unchanged [21]. The term *narrowband* is used under the assumption, satisfied in most of the cases, of a slowing varying signal envelope when either the signals' or the sensor elements' bandwidth is small relative to the frequency of operation. This assumption can be also extended to wideband signals, provided the frequency response of the array is approximately flat over the signals' bandwidth and the propagation time across the array is small compared to the reciprocal bandwidths. Under this model, the received signals can be expressed as a superposition of signals from all the sources and linearly added noise represented by

$$\mathbf{x}(t) = \sum_{k=1}^{K} \mathbf{a}(\theta_k) s_k(t) + \mathbf{n}(t) \qquad (5.5)$$

where $\mathbf{x}(t) \in \mathbb{C}^N$ is the complex baseband equivalent received signal vector at the antenna array at time t, or

$$\mathbf{x}(t) = [x_1(t), x_2(t), \ldots, x_N(t)]^T, \qquad (5.6)$$

$s_k(t)$ is the incoming plane wave from the kth source at time t and arriving from the direction θ_k, $\mathbf{a}(\theta_k) \in \mathbb{C}^N$ is the array response vector to this direction, and $\mathbf{n}(t) \in \mathbb{C}^N$ represents additive noise. Note that whatever appears in the complex vector $\mathbf{n}(t)$ is the noise either "sensed" along with the signals or generated internal to the instrumentation [126]. A single observation $\mathbf{x}(t)$ from the array is often referred to as a *snapshot*. In matrix notation, (5.5) can be written as

$$\mathbf{x}(t) = \mathbf{A}(\mathbf{\Theta}) \mathbf{s}(t) + \mathbf{n}(t) \qquad (5.7)$$

where $\mathbf{A}(\mathbf{\Theta}) \in \mathbb{C}^{N \times K}$ is the array response matrix parameterized by the direction of arrival (DOA) (i.e. each column of which represents the array response vector for each signal source), or

$$\mathbf{A}(\mathbf{\Theta}) = [\mathbf{a}(\theta_1), \mathbf{a}(\theta_2), \ldots, \mathbf{a}(\theta_K)], \qquad (5.8)$$

$\mathbf{\Theta}$ is the vector of all the DOAs, or

$$\mathbf{\Theta} = [\theta_1, \theta_2, \ldots, \theta_K]^T \qquad (5.9)$$

and $\mathbf{s}(t) \in \mathbb{C}^K$ represents the vector of the incoming signal in amplitude and phase from each signal source at time t, or

$$\mathbf{s}(t) = [s_1(t), s_2(t), \ldots, s_K(t)]^T. \qquad (5.10)$$

Usually, $\mathbf{s}(t)$ is referred to as the *desired signal portion* of $\mathbf{x}(t)$. The three most important features of (5.7) are that the matrix $\mathbf{A}(\boldsymbol{\Theta})$ must be time-invariant over the observation interval, the model is bilinear in $\mathbf{A}(\boldsymbol{\Theta})$ and $\mathbf{s}(t)$, and the noise is additive [21].

The set of array response vectors corresponding to all possible directions of arrival in (5.7), $\mathbf{A}(\boldsymbol{\Theta})$, is also referred to as the *array manifold* (AM). In simple words, each element a_{ij} ($i = 1, 2, \ldots, N, j = 1, 2, \ldots, K$) of the AM, $\mathbf{A}(\boldsymbol{\Theta})$, indicates the response of the ith element to a signal incident from the direction of the jth signal. The majority of algorithms developed for the estimate of the DOAs require that the array response matrix $\mathbf{A}(\boldsymbol{\Theta})$ be completely known for a given parameter vector $\boldsymbol{\Theta}$ [127]. This is usually accomplished by direct calibration in the field, or by analytical means using information about the position and response of each individual sensor (such as is done with a *uniform linear array*, for example).

An *unambiguous* array manifold $\mathbf{A}(\boldsymbol{\Theta})$ is defined to be one which any collection of $K \leq N$ distinct vectors from $\mathbf{A}(\boldsymbol{\Theta})$ forms a linearly independent set. For example, an element from the array manifold (an array response vector for a single signal source) of a uniform linear array of identical sensors, as shown in Fig. 5.2, is proportional to

$$\mathbf{a}(\theta_k) = \begin{bmatrix} 1 \\ e^{j\frac{2\pi}{\lambda} d \sin \theta_k} \\ e^{j\frac{2\pi}{\lambda} 2d \sin \theta_k} \\ \vdots \\ e^{j\frac{2\pi}{\lambda}(N-1)d \sin \theta_k} \end{bmatrix} \qquad (5.11)$$

where λ is the wavelength of the impinging wavefront and d is the distance between adjacent elements. For a range of angles of arrival $\theta \in \left[-\frac{\pi}{2}, \frac{\pi}{2}\right]$ (meaningful for the particular geometry), it is obvious that the AM maintains its unambiguity provided $d < \frac{\lambda}{2}$. In the case that $\theta_{\max} < \frac{\pi}{2}$ is the maximum bearing deviation from broadside that is expected or imposed by operational considerations, then the wavefield must be sampled at a rate such that $d < \frac{\lambda}{2} \frac{1}{\sin \theta_{\max}}$. For more widely spaced sensors, it is possible that there may exist pairs of angles θ_i and θ_j, with $\theta_i \neq \theta_j$, such that $\mathbf{a}(\theta_i) = \mathbf{a}(\theta_j)$. This equality holds when $\frac{d}{\lambda} \sin \theta_i = n + \frac{d}{\lambda} \sin \theta_j$, where $n \in \mathbb{Z}, n \neq 0$. In such cases, the array response for a signal arriving from angle θ_i is indistinguishable from that arriving from angle θ_j.

Uniform sampling of the wavefield implies that all the lags are sampled at least once, and hence, no ambiguous locations should result since the correlation function is completely known [125]. Even though the sampling structure leads to a convenient method of computing a beamformed output by exploiting a structure amenable to FFT processing, it does not need to be uniform [125]. In fact, there may exist cases that it is not required or desirable. Note at this point that the requirement for the interelement spacing in a uniform linear array to be less than half of the wavelength of the highest frequency in the receiver band can be interpreted as the spatial analog to the well-known Nyquist sampling criterion which allows the reconstruction of a continuous-time wavefront occupying a bandwidth B from its discrete-time samples if these are taken with sampling frequency of not less than $2B$. If $\mathbf{A}(\Theta)$ is unambiguous and $N \geq K$, then $\mathbf{A}(\Theta)$ will be of full-rank K. In a similar manner, for an array manifold with resolved ambiguity, knowing the mode vector $\mathbf{a}(\theta_i)$ is tantamount to knowing the angle θ_i [126].

Furthermore, for a set of data observations $L > K$, we can form the matrices

$$\mathbf{X} = [\mathbf{x}(1), \mathbf{x}(2), \ldots, \mathbf{x}(L)], \tag{5.12a}$$
$$\mathbf{S} = [\mathbf{s}(1), \mathbf{s}(2), \ldots, \mathbf{s}(L)], \text{ and} \tag{5.12b}$$
$$\mathbf{N} = [\mathbf{n}(1), \mathbf{n}(2), \ldots, \mathbf{n}(L)] \tag{5.12c}$$

where $\mathbf{X}$ and $\mathbf{N} \in \mathbb{C}^{N \times L}$ and $\mathbf{S} \in \mathbb{C}^{K \times L}$, and further write

$$\mathbf{X} = \mathbf{A}(\Theta)\,\mathbf{S} + \mathbf{N}. \tag{5.13}$$

5.4   THE SUBSPACE-BASED DATA MODEL

Ignoring the noise effects in (5.13), each observation of the received signal, $\mathbf{A}(\Theta)\,\mathbf{S}$, is constrained to lie in the K-dimensional subspace $\mathbb{C}^N$ defined by the K columns of $\mathbf{A}(\Theta)$.

Fig. 5.3 illustrates this idea for the special case of two sources ($K = 2$) and four snapshots ($L = 4$). Each of the two sources has associated with it a response vector $\mathbf{a}(\theta_k)$ from the array manifold, and the four snapshots $\mathbf{x}(t_1), \ldots, \mathbf{x}(t_4)$ lie in the two-dimensional subspace spanned by these vectors. The specific positions of these vectors depend on the signal waveforms at each time instant. Note that the array manifold intersects the signal subspace at only two points, each corresponding to a response of one of the signals [21].

Even though $L > K$, it is possible, however, for the signal subspace to have dimension smaller than K. This occurs if the matrix of signal samples $\mathbf{S}$ has a rank less than K. This situation may arise, for example, if one of the signals is a linear combination of the others. Such signals are referred to as *coherent* or fully-correlated signals, and occur most frequently in the sensor array problem in a *multipath propagation* scenario. Multipath results when a given signal is received at the array from several different directions or paths due to reflections from various objects in the wireless channel. It may also be possible that the available snapshots are fewer than the emitting sources, in which case the signal subspace cannot exceed the number of observations [21]. In either case, the dimension of the signal subspace is less than the number of present sources. However, this does not imply that estimates of the number of sources are impossible. For instance, it can be shown [126] that for one-parameter vectors, the angle of arrival in our case (or any other one parameter per source), the signal parameters are still identifiable if $\mathbf{A}(\Theta)$ is unambiguous and $N > 2K - K'$, where $K' = \text{rank}\,[\mathbf{A}(\Theta)\,\mathbf{S}]$.

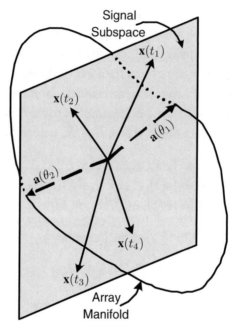

FIGURE 5.3: A geometric view of the DOA estimation problem [21].

The identifiability condition, geometrically obvious, is that the signal subspace be spanned by a *unique* set of K vectors from the array manifold.

In the event that the measurements made are more than the present signals (i.e., the number of sources K is less than the number of elements N), the data model in (5.7) admits an appealing geometric interpretation and provides insight into the sensor array processing problem [21]. The measurements taken form the vectors of complex values with dimension in space equal to the number of elements in the array (N). In the absence of noise, the expression which gives $\mathbf{x}(t)$ in (5.7), $\mathbf{A}(\boldsymbol{\Theta})\mathbf{s}(t)$, is confined to a space dimension K' (at most a K-dimensional subspace of $\mathbb{C}^N$), referred to as the *signal subspace* and it spans either the entire or some fraction of the column space of $\mathbf{A}(\boldsymbol{\Theta})$. If any of the impinging signals are perfectly correlated, i.e., one signal is simply complex scalar multiple of another, the span of the *signal subspace* K' will be less than K. Consequently, if there is sufficient *excitation*, in other words no signals are *perfectly correlated*, the signal subspace is K-dimensional. Considering noise, since it is typically assumed to possess energy in all dimensions of the observation space, (5.7) is often referred to as a *low-rank signal* in *full-rank noise* data model.

This entire geometric picture leads to the accurate parameter estimation problem by handling it as subspace intersection. Because of the many applications for which the subspace-based data method is appropriate, numerous *subspace-based* techniques have been developed to exploit it [21].

5.5 SIGNAL AUTOCOVARIANCE MATRICES

Before we discuss the algorithms for DOA estimation, we first need to define two commonly used terms: the received signal autocovariance matrix $\mathbf{R}_{xx}$ and the desired signal autocovariance matrix $\mathbf{R}_{ss}$ given by

$$\mathbf{R}_{xx} = \mathcal{E}\left\{\mathbf{x}(t)\mathbf{x}^H(t)\right\} \tag{5.14}$$
$$\mathbf{R}_{ss} = \mathcal{E}\left\{\mathbf{s}(t)\mathbf{s}^H(t)\right\} \tag{5.15}$$

where H denotes Hermitian (or complex-conjugate transpose) matrix operation and $\mathcal{E}\{\cdot\}$ is the expectation operation on the argument. In reality, the expected value cannot be obtained exactly since an infinite time interval is necessary and estimates, as the average over a finite, sufficiently enough, number of data "snapshots" must be used in practical implementations as

$$\hat{\mathbf{R}}_{xx} \triangleq \lim_{M \to \infty} \frac{1}{M} \sum_{m=1}^{M} \mathbf{x}(t_m)\mathbf{x}^H(t_m). \tag{5.16}$$

The same approximation holds for $\hat{\mathbf{R}}_{ss}$. With the typical assumption that the incident signals are noncoherent, the source covariance matrix $\mathbf{R}_{ss}$ is positive definite [128]. In addition, the noise is typically assumed to be a complex stationary *Gaussian random process*. The motivation for this assumption is that if there are many sources of noise, the sum will be Gaussian distributed according to the *central limit theorem* [129]. Also, further analysis of direction finding performance is greatly simplified by assuming white Gaussian noise.

If, additionally, it is assumed to be uncorrelated both with the signals, and for successive signal samples, (5.14) can be written as

$$\begin{aligned}\mathbf{R}_{xx} &= \mathbf{A}(\Theta)\,\mathbf{R}_{ss}\mathbf{A}^H(\Theta) + \mathcal{E}\left\{\mathbf{n}(t)\mathbf{n}^H(t)\right\} \\ &= \mathbf{A}(\Theta)\,\mathbf{R}_{ss}\mathbf{A}^H(\Theta) + \sigma_n^2\Sigma\end{aligned} \tag{5.17}$$

where σ_n^2 is the noise variance and Σ is normalized so that $\det(\Sigma) = 1$. The simplifying assumption of spatial whiteness (i.e., $\Sigma = \mathbf{I}$, where $\mathbf{I}$ is the identity matrix) is often made.

The assumptions of a known array response and known noise covariance are never practically valid. Due to changes in the weather, reflective and absorptive bodies in the nearby surrounding environment, and antenna location, the response of the array may be substantially different than it was last calibrated [130]. Furthermore, the calibration measurements themselves are subject to gain and phase errors. For the case of analytically calibrated arrays of identical elements, including orientation, errors may occur because the elements are not really identical and their locations are not precisely known. Depending on the degree to which the actual antenna response differs from its nominal value, the performance of a particular algorithm may significantly be degraded [130].

Since the surrounding environment of the array may be time-varying, the requirement of known noise statistics is also difficult to satisfy in practice. In addition, effects of unmodeled "noise" phenomena such as distributed sources, reverberation, noise due to the antenna platform, and undesirable channel crosstalk are often unable to be accounted for. Measurement of the noise statistics is usually a complicated task due to the fact that signals-of-interest are often observed along with the noise and interference. When signal subspace methods are applied for DOA estimation, it is often assumed that the noise field is isotropic, independent from channel to channel and equal at each one [130], which is not the case in reality. For high signal-to-noise (SNR) ratio, deviations of the noise from these assumptions are not critical since they contribute little to the statistics of the received by the array signal. However, at low SNR values, the degradation in the algorithms' performance may be severe.

5.6   CONVENTIONAL DOA ESTIMATION METHODS

Two methods are usually classified as conventional methods: the *Conventional Beamforming Method* and *Capon's minimum Variance Method* [13].

5.6.1   Conventional Beamforming Method

The *conventional beamforming method* (CBF) is also referred to as the delay-and-sum method or Bartlett method. The idea is to scan across the angular region of interest (usually in discrete steps), and whichever direction produces the largest output power is the estimate of the desired signal's direction. More specifically, as the look direction θ is varied incrementally across the space of access, the array response vector $\mathbf{a}(\theta)$ is calculated and the output power of the beamformer is measured by

$$P_{CBF}(\theta) = \frac{\mathbf{a}^H(\theta)\mathbf{R}_{xx}\mathbf{a}(\theta)}{\mathbf{a}^H(\theta)\mathbf{a}(\theta)}. \qquad (5.18)$$

This quantity is also referred to as the *spatial spectrum* and the estimate of the true DOA is the angle θ that corresponds to the peak value of the output power spectrum.

The method is also referred to as *Fourier method* since it is a natural extension of the classical Fourier based spectral analysis with different window functions [131, 132]. In fact, if a ULA of isotropic elements is used, the spatial spectrum in (5.18) is a spatial analog of the classical periodogram in time-series analysis. Note that other types of arrays correspond to nonuniform sampling schemes in time-series analysis. As with the periodogram, the spatial spectrum has a resolution threshold. That is, an array with only a few elements is not able to form neither narrow nor sharp peaks and hence, its ability to resolve closely spaced signals sources is limited [13]. More accurately, waves arriving with electrical angle separation[1] less than $2\pi/N$ cannot be resolved with this method. For example, using a five-element ULA with an element separation of $d = \lambda/2$ results in a resolution threshold of $23°$ [133]. The poor resolution is a significant weakness of the method. Other choices of weighting vectors **w**, that result in lower resolution thresholds, have been therefore investigated.

5.6.2   Capon's Minimum Variance Method

The Capon's minimum variance method is also known as the *minimum variance distortionless look* (MVDL). The MVDL is an attempt to overcome the poor resolution problem associated with the delay-and-sum method and it results a significant improvement [17]. In this method, the output power is minimized with the constraint that the gain in the desired direction remains unity. Solving this constraint optimization problem for the weight vector [13, 134] we obtain

$$\mathbf{w} = \frac{\mathbf{R}_{xx}^{-1}\mathbf{a}(\theta)}{\mathbf{a}^H(\theta)\mathbf{R}_{xx}^{-1}\mathbf{a}(\theta)} \tag{5.19}$$

which gives the *Capon's Spatial Spectrum*:

$$P_{\text{Capon}}(\theta) = \mathbf{w}^H\mathbf{R}_{xx}\mathbf{w} = \frac{1}{\mathbf{a}^H(\theta)\mathbf{R}_{xx}^{-1}\mathbf{a}(\theta)}. \tag{5.20}$$

Again, the estimate of the true direction of arrival is the angle θ that corresponds to the peak value in this spectrum. The MVDL only requires an additional matrix inversion compared to the CBF and exhibits greater resolution in most cases.

In general, the conventional DOA estimation algorithms provide some important advantages. Computing the spatial power spectrum for one range of θ does not prevent the algorithm from subsequently computing the spectrum for another range of θ using the same data. The spatial characteristics of the data for all directions are compactly represented by $\mathbf{R}_{xx}$, and they are needed to be computed only once. Thus, the method does not have blind spots in time during which transient signals, away from directions of constantly transmitting sources, can appear intermittently and fail to be detected [134]. Another advantage is that by steering the antenna electronically rather than mechanically, the speed of the scan through a region of interest is limited by computational speed instead of mechanical speed.

5.7   SUBSPACE APPROACH TO DOA ESTIMATION

The other main group of DOA estimation algorithms are called the *subspace methods*. Geometrically, the received signal vectors form the received signal vector space whose vector dimension

---

[1]The electrical angle for a ULA is defined as $kd\sin\theta$.

is equal to the number of array elements N. The received signal space can be separated into two parts: the *signal subspace* and the *noise subspace*. The *signal subspace* is the subspace spanned by the columns of $\mathbf{A}(\Theta)$ [21], and the subspace orthogonal to the signal subspace is known as the *noise subspace*. The subspace algorithms exploit this orthogonality to estimate the signals' DOAs.

5.7.1   The MUSIC Algorithm

Within the class of the so-called *signal-subspace* algorithms, MUSIC (*MUltiple SIgnal Classification*)[123, 135] has been the most widely examined. In a detailed performance evaluation based on hundreds of simulations, MIT's Lincoln Laboratories concluded that, among the high-resolution algorithms then available, MUSIC was the most promising and a leading candidate for further study and actual hardware implementation [136].

The popularity of the MUSIC algorithm is in large part due to its generality. For example, it is applicable to arrays of arbitrary but known configuration and response, and can be used to estimate multiple parameters per source (e.g., azimuth, elevation, range, polarization, etc.). However, this generality is accompanied with the expense that the array response must be known for all possible combinations of source parameters; i.e., the response must be either measured (*calibrated*) and stored, or one must be able to characterize it analytically (e.g., as in the case of root-MUSIC [123, 137]). In addition, MUSIC requires *a priori* knowledge of the second-order spatial statistics of the background noise and interference field. These assumptions are never satisfied in reality as explained earlier.

The MUSIC algorithm was developed by Schmidt [123, 126] by noting that the desired signal array response is orthogonal to the noise subspace. The signal and noise subspaces are first identified using eigendecomposition of the received signal covariance matrix. Following, the *MUSIC spatial spectrum* is computed, from which the DOAs are estimated. Inside the algorithm, we first define the general *array manifold* to be the set

$$\mathcal{A} = \big\{ \mathbf{a}(\theta_i) : \theta_i \in \Theta \big\} \tag{5.21}$$

for some region Θ of interest in the DOA space. The array manifold is assumed unambiguous and known for all the values of angle θ, either analytically or through some calibration procedure. The objective is to apply appropriate methods to the received signals so as to extract the region θ out of the range of Θ.

If noise was absent in (5.7), the observations $\mathbf{x}(t)$ would be confined entirely to the K-dimensional subspace of $\mathbb{C}^K$ defined by the span of $\mathbf{A}(\Theta)$. Determining the DOAs for the no-noise case is simply a matter of finding the K unique elements of $\mathcal{A}$ that intersect this subspace [130]. A different approach is necessary in the presence of noise since the observations become "full-rank". The approach of MUSIC, and other subspace-based methods, is to first estimate the dominant subspace of the observations, and then find the elements of $\mathcal{A}$ that are in some sense closest to this subspace.

The subspace estimation step is typically achieved by eigendecomposition of the autoco-variance matrix of the received data $\mathbf{R}_{xx}$. For MUSIC to be applicable, the emitter covariance $\mathbf{R}_{ss}$ is required to be full-rank, i.e., that $K' = K$. Using the model in (5.17) and assuming spatial whiteness[2], i.e., $\mathcal{E}\left\{\mathbf{n}(t)\mathbf{n}^H(t)\right\} = \sigma_n^2 \mathbf{I}$, the eigendecomposition of $\mathbf{R}_{xx}$ will give the eigenvalues λ_n such that $\lambda_1 > \lambda_2 > \ldots > \lambda_K > \lambda_{K+1} = \lambda_{K+2} = \ldots = \lambda_N = \sigma_n^2$ and the corresponding eigenvectors $\mathbf{e}_n \in \mathbb{C}^N$, $n = 1, 2, \ldots, N$, of $\mathbf{R}_{xx}$. Furthermore, it is easily shown that $\mathbf{R}_{xx}$ can be written in the following form [138]:

$$\mathbf{R}_{xx} = \sum_{n=1}^{N} \lambda_n \mathbf{e}_n \mathbf{e}_n^H = \mathbf{E}\boldsymbol{\Lambda}\mathbf{E}^H = \mathbf{E}_s \boldsymbol{\Lambda}_s \mathbf{E}_s^H + \mathbf{E}_n \boldsymbol{\Lambda}_n \mathbf{E}_n^H$$

$$= \mathbf{E}_s \boldsymbol{\Lambda}_s \mathbf{E}_s^H + \sigma_n^2 \mathbf{E}_n \mathbf{E}_n^H = \mathbf{E}_s \tilde{\boldsymbol{\Lambda}}_s \mathbf{E}_s^H + \sigma_n^2 \mathbf{I} \tag{5.22}$$

where $\mathbf{E} = [\mathbf{e}_1, \mathbf{e}_2, \ldots, \mathbf{e}_N]$, $\mathbf{E}_s = [\mathbf{e}_1, \mathbf{e}_2, \ldots, \mathbf{e}_K]$, $\mathbf{E}_n = [\mathbf{e}_{K+1}, \mathbf{e}_{K+2}, \ldots, \mathbf{e}_N]$, $\boldsymbol{\Lambda} = \mathrm{diag}\{\lambda_1, \lambda_2, \ldots, \lambda_N\}$, $\boldsymbol{\Lambda}_s = \mathrm{diag}\{\lambda_1, \lambda_2, \ldots, \lambda_K\}$, $\boldsymbol{\Lambda}_n = \mathrm{diag}\{\lambda_{K+1}, \lambda_{K+2}, \ldots, \lambda_N\}$, and $\tilde{\boldsymbol{\Lambda}}_s = \boldsymbol{\Lambda}_s - \sigma_n^2 \mathbf{I}$. The eigenvectors $\mathbf{E} = [\mathbf{E}_s, \mathbf{E}_n]$ can be assumed to form an orthonormal basis (i.e., $\mathbf{E}\mathbf{E}^H = \mathbf{E}^H\mathbf{E} = \mathbf{I}$). The span of the K vectors $\mathbf{E}_s$ defines the *signal* subspace, and the orthogonal complement spanned by $\mathbf{E}_n$ defines the noise *subspace*. For a detailed analysis of the eigenstructure properties of the signal autocovariance matrices $\mathbf{R}_{xx}$ and $\mathbf{R}_{ss}$ the reader is referred to [126]. Once the subspaces are determined, the DOAs of the desired signals can be estimated by calculating the MUSIC spatial spectrum over the region of interest [21]:

$$P_{\mathrm{MUSIC}}(\theta) = \frac{\mathbf{a}^H(\theta)\mathbf{a}(\theta)}{\mathbf{a}^H(\theta)\mathbf{E}_n \mathbf{E}_n^H \mathbf{a}(\theta)}. \tag{5.23}$$

Note that the $\mathbf{a}(\theta)$s are the array response vectors calculated for all angles θ within the range of interest. Because the desired array response vectors $\mathbf{A}(\boldsymbol{\Theta})$ are orthogonal to the noise subspace, the peaks in the MUSIC spatial spectrum represent the DOA estimates for the desired signals. Due to imperfections in deriving $\mathbf{R}_{xx}$, the noise subspace eigenvalues will not be exactly equal to σ_n^2. They do, however, form a group around the value σ_n^2 and can be distinguished from the signal subspace eigenvalues. The separation becomes more pronounced as the number of samples used in the estimation of $\mathbf{R}_{xx}$ increases (ideally reaches infinity).

To demonstrate the efficiency of the algorithm, we choose as an example a ULA with $N = 8$ and $d = \lambda/2$. We assume four equal-power uncorrelated sources ($K = 4$) located in the *far-field* of the array, with $\theta_1 = +60°$, $\theta_2 = +15°$, $\theta_3 = -30°$, and $\theta_4 = -75°$. Moreover,

---

[2]The assumption of spatially white noise is not necessary; the extension to an arbitrary noise autocovariance $\sigma_n^2 = \boldsymbol{\Sigma}$ is straightforward, provided that $\boldsymbol{\Sigma}$ is known.

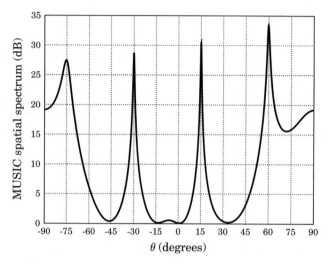

FIGURE 5.4: Spatial spectrum of the MUSIC algorithm.

uncorrelated *spatially white Gaussian noise* with zero mean and unit variance ($\sigma_n^2 = 1$) is assumed. A total of 500 observations are taken ($L = 500$). Fig. 5.4 displays the obtained MUSIC spatial spectrum. The performance of the algorithm is shown to be excellent, since the peaks in the spatial spectrum are located at angles being exactly the DOAs.

A final remark for the algorithm's performance is that MUSIC yields asymptotically unbiased parameter estimates, even for multiple incident wavefronts, because both $\mathbf{R}_{ss}$ and $\mathbf{E}_n$ are asymptotically perfectly measured [139].

5.7.2   The ESPRIT Algorithm

Although the performance advantages of MUSIC are substantial, they are achieved at a considerable cost in computation (searching over parameter space) and storage (of array calibration data). Moreover, even for the *one-dimensional* MUSIC estimation (DOA in the particular case), there exist several drawbacks although being conceptually easy. Primarily, problems in the finite measurement case arise from the fact that since K signals are known to be present, the search for their DOAs, $(\theta_1, \theta_2, \ldots, \theta_K)$, should be sought simultaneously by maximizing an appropriate functional rather than obtaining estimates one at a time as is done in the search for spectral peaks over $P_{MUSIC}(\theta)$. However, multidimensional searches are accompanied with an intense expense compared to one-dimensional searches. The reduction in computational load achieved with an *one-dimensional* search for K parameters comes with the trade-off of the method being finite-sample-biased in a multisource environment [122]. Furthermore, in either low SNR scenarios or closely spaced sources (i.e., multiple peaks observed in the measurements) MUSIC's performance reduces dramatically. Nevertheless, despite its drawbacks, it should be emphasized that MUSIC has proven to outperform techniques existed prior to its development [136].

ESPRIT (*Estimation of Signal Parameters via Rotational Invariant Techniques*) is similar to MUSIC in that it correctly exploits the underlying data model. Beyond retaining most of the essential features of the *arbitrary* array of sensors, ESPRIT achieves a significant reduction in the aforementioned computation and storage costs. This is done by imposing a constraint on the structure of the sensor array to possess a *displacement invariance*, i.e., sensors occur in matched pairs with identical displacement vectors [122]. Such conditions, are or can be satisfied in many practical problems. In addition to obtaining signal parameters efficiently, ESPRIT is also less sensitive to array imperfections than other techniques including MUSIC [137]. The discussions herein will be focused on the DOA estimation problem, although ESPRIT is generally applicable to a wide variety of problems. The method *simultaneously* estimates the number of sources and DOAs [140].

The majority of the algorithms developed for the DOA estimation problem require that the array response matrix $\mathbf{A}(\mathbf{\Theta})$ be completely known for a given parameter vector $\mathbf{\Theta}$. This is usually accomplished by either analytical means using information about the position and response of each individual sensor (such as is done with a ULA, for example) or direct calibration in the field. A notable expression is the ESPRIT algorithm [140, 141], which assumes, instead, that the N-element array is composed of two identical translated N'-element subarrays, where $N' < N \leq 2N'$, as depicted in Fig. 5.5. The individual elements of each subarray may have arbitrary directional gain and phase responses, provided that each one has an identical twin in the companion subarray [124]. The elements in each pair of identical sensors, or *doublet*, are assumed to be separated by a fixed displacement vector $\mathbf{D}$. For certain special array configurations, the subarrays may *overlap*, i.e., an array element may be a member of both subarrays ($N < 2N'$) as shown in Fig. 5.5(a). For subarrays that do not share elements, $N = 2N'$, as shown in Fig. 5.5(b).

The ESPRIT assumption of rotationally invariant subarrays leads to a very special form of $\mathbf{A}(\boldsymbol{\theta})$. Employing the configuration shown in Fig. 5.5, the output of the array is modeled as

$$\mathbf{x}(t) = \begin{bmatrix} \mathbf{A}_1(\mathbf{\Theta}) \\ \mathbf{A}_2(\mathbf{\Theta}) \end{bmatrix} \mathbf{s}(t) + \begin{bmatrix} \mathbf{n}_1(t) \\ \mathbf{n}_2(t) \end{bmatrix} \tag{5.24}$$

where $\mathbf{A}_1(\mathbf{\Theta})$, $\mathbf{A}_2(\mathbf{\Theta}) \in \mathbb{C}^{N \times K}$ indicate the array manifold of each subarray, respectively, and $\mathbf{n}_1(t)$, $\mathbf{n}_2(t) \in \mathbb{C}^{N \times 1}$ represent the noise collected by each subarray, respectively. Furthermore,

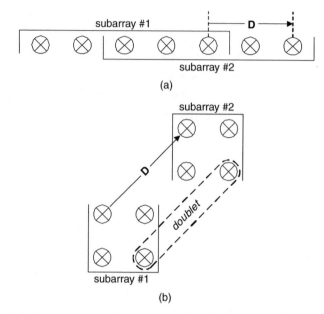

FIGURE 5.5: ESPRIT sensor array geometry: (a) One array consists of two overlapping arrays, whereas the other (b) consists of two identical and disjoint arrays [140].

if we let $\mathbf{J}_1$ and $\mathbf{J}_2$ represent the $N' \times N$ selection matrices that assign the elements of the entire array to each of the two subarrays as

$$\mathbf{J}_1 = \left[\mathbf{I}_{N'} \vdots \mathbf{0}_{N' \times (N-N')} \right] \qquad (5.25a)$$

$$\mathbf{J}_2 = \left[\mathbf{0}_{N' \times (N-N')} \vdots \mathbf{I}_{N'} \right] \qquad (5.25b)$$

where $\mathbf{I}_{N'}$ is the $N' \times N'$ identity matrix and $\mathbf{0}_{N' \times (N-N')}$ is the $N' \times (N - N')$ matrix of zeros, it is easy to see that an array composed of two identical subarrays satisfies [142]

$$\mathbf{J}\mathbf{A}(\mathbf{\Theta}) = \begin{bmatrix} \mathbf{J}_1 \\ \mathbf{J}_2 \end{bmatrix} \mathbf{A}(\mathbf{\Theta}) = \begin{bmatrix} \mathbf{A}_1(\mathbf{\Theta}) \\ \mathbf{A}_1(\mathbf{\Theta})\mathbf{\Phi} \end{bmatrix} \qquad (5.26)$$

where $\mathbf{\Phi}$ is a unitary diagonal matrix with diagonal elements ϕ_i given by [142]

$$\phi_i = \exp\left\{-j\boldsymbol{\beta}_i^T \cdot \mathbf{D}\right\}, \quad i = 1, 2, \dots, K \qquad (5.27)$$

where $\boldsymbol{\beta}_i$ is the vector wavenumber of the incident plane from the ith narrowband source and $\mathbf{D}$ is the vector displacement between the two subarrays. If we assume the total array to be linear and the orientation of $\mathbf{D}$ to be toward $\frac{\pi}{2}$ (rather than $-\frac{\pi}{2}$), as shown in Fig. 5.5(a), $\boldsymbol{\beta}_i \cdot \mathbf{D}$ simplifies to $-\frac{2\pi}{\lambda} \sin \theta_i$, where λ is the wavelength of the narrowband signal, with θ_i being the angle of arrival from the ith source. As it is obvious from (5.26), ESPRIT does not exploit the entire array manifold. The knowledge that is used, and consequently required, is the response of one subarray and the displacement structure of the array. Since $\mathbf{A}_1(\boldsymbol{\Theta})$ must be full-rank ($K \leq N'$ for all $\boldsymbol{\Theta}$), the resolvable sources when applying the ESPRIT algorithm are limited to N'.

ESPRIT exploits the structure of (5.26) in the following way. If $\mathbf{E}_s \in \mathbb{C}^{N \times K}$ represents the eigenvectors corresponding to the K largest eigenvalues of the received signal autocovariance matrix $\mathbf{R}_{xx}$, and if no pairs of signals are correlated, then it is easily shown that [142]

$$\mathbf{E}_s \triangleq \begin{bmatrix} \mathbf{E}_1 \\ \mathbf{E}_2 \end{bmatrix} = \begin{bmatrix} \mathbf{A}_1(\boldsymbol{\Theta}) \\ \mathbf{A}_1(\boldsymbol{\Theta})\boldsymbol{\Phi} \end{bmatrix} \mathbf{T} \tag{5.28}$$

for some full-rank matrix $\mathbf{T} \in \mathbb{C}^{K \times K}$. Solving for $\mathbf{A}_1(\boldsymbol{\Theta})$ and substituting into the lower block of (5.28), leads to [142]

$$\mathbf{E}_2 = \mathbf{E}_1 \mathbf{T}^{-1} \boldsymbol{\Phi} \mathbf{T} = \mathbf{E}_1 \boldsymbol{\Psi} \tag{5.29}$$

where the matrix $\boldsymbol{\Psi} = \mathbf{T}^{-1}\boldsymbol{\Phi}\mathbf{T}$ has been defined (or $\boldsymbol{\Phi} = \mathbf{T}\boldsymbol{\Psi}\mathbf{T}^{-1}$). Thus, since $\boldsymbol{\Phi}$ and $\boldsymbol{\Psi}$ are related by a similarity transformation, the eigenvalues of $\boldsymbol{\Psi}$ must be equal to the diagonal elements of $\boldsymbol{\Phi}$. Furthermore, the columns of $\mathbf{T}$ are the eigenvectors of $\boldsymbol{\Phi}$ [122]. This is the fundamental relationship in the development of ESPRIT and its properties. Consequently, if $N' \geq K$ and $D = |\mathbf{D}| < \frac{\lambda}{2}$, the DOAs may be uniquely determined from the eigenvalues of the operator $\boldsymbol{\Psi}$ that maps $\mathbf{E}_1$ onto $\mathbf{E}_2$ as

$$\theta_k = \sin^{-1} \left\{ \frac{\arg\{\psi_i\}}{\frac{2\pi}{\lambda} D} \right\}, \quad i = 0, 1, \ldots, K \tag{5.30}$$

where ψ_i represents each of the eigenvalues of $\boldsymbol{\Psi}$. Note that this result is independent of the actual value $\mathbf{A}(\boldsymbol{\Theta})$ (as long as remains full-rank) and, thus, the array needs not to be calibrated in order to estimate the DOAs [124].

5.8   UNIQUENESS OF DOA ESTIMATES

Given a number of uncorrelated signals less than the number of sensors and an unlimited supply of data, most of the preceding DF methods can uniquely and exactly locate the sources. However, in the presence of too many signals, or the availability of only a finite amount of data, any given DF algorithm can yield erroneous DOA estimates or fail completely [134]. In [143], Wax and Ziskind derive a maximum in the number of present signals such that the DOAs can be estimated uniquely. They show that certain conditions on the array manifold, the number of sensors, the number of signals, and the rank of the autocovariance matrix of the received signals, determine whether or not the DOAs of the signals can be estimated uniquely. Depending on the strength of the conditions, uniqueness can be either guaranteed for every possible batch of received data or assured with unitary probability. Both cases require that the array manifold be known and that the array response vectors corresponding to maximum N distinct DOAs be linearly independent for all those choices of DOAs.

The strongest condition states that uniqueness is guaranteed if the number K of signals is less than the average of the number N of sensors and the rank of the signal autocorrelation matrix [143]:

$$K < \frac{N + \text{rank}\left\{\mathbf{R}_{ss}\right\}}{2}. \tag{5.31}$$

For example, if the signals are uncorrelated and rank rank$\left\{\mathbf{R}_{ss}\right\} = N$, then (5.31) merely states the familiar condition $K < N$. However, the effect of correlated signals is to reduce the rank of $\mathbf{R}_{ss}$ and consequently to reduce the maximum number of sources that can be localized uniquely. For example, if all signals are fully-correlated, i.e., rank$\left\{\mathbf{R}_{ss}\right\} = 1$, the number of uniquely localized sources reduces to $L < (M+1)/2$. The strong condition in (5.31) seems to be very restrictive in the case of a multipath environment. If uniqueness with probability one can be accepted, rather than guaranteed uniqueness, the following weaker condition is sufficient [143]:

$$K < \frac{2\text{rank}\left\{\mathbf{R}_{ss}\right\}}{2\text{rank}\left\{\mathbf{R}_{ss}\right\} + 1} M. \tag{5.32}$$

This condition reduces to $K < N$ for uncorrelated signals, as well, but for fully-correlated signals reduces to the improved limit $K < \frac{2}{3}N$.

CHAPTER 6

Beamforming Fundamentals

With the direction of the incoming signals known or estimated, the next step is to use spatial processing techniques to improve the reception performance of the receiving antenna array based on this information. Some of these spatial processing techniques are referred to as *beamforming* because they can form the array beampattern to meet the requirements dictated by the wireless system. Given a 1D linear array of elements and an impinging wavefront from an arbitrary point source, the directional power pattern $P(\theta)$ can be expressed as [59, 125]

$$P(\theta) = \int a(x)e^{-j\beta d(x,\theta)}dx \qquad (6.1)$$

where $a(x)$ is the amplitude distribution along the array, β is the phase constant, and $d(x, \theta)$ is the relative distance the impinging wavefront, with an angle of arrival θ, has to travel between points uniformly spaced a distance x apart along the length of the array. The exponential term is the one that primarily scans the beam of the array in a given angular direction. The integral of (6.1) can be generalized for two- and three-dimensional configurations [59]. Equation (6.1) is basically the Fourier transform of $a(x)$ along the length of the array and is the basis for beamforming methods [125]. The amplitude distribution $a(x)$, necessary for a desired $P(\theta)$, is usually difficult to implement practically [59]. Therefore, realization of (6.1) most of the times is accomplished using discrete sources, represented by a summation over a finite number of elements [59]. Thus, by controlling the relative phase between the elements, the beam can be scanned electronically with some possible changes in the overall shape of the array pattern. This is the basic principle of array phasing and beam shaping.

The main objective of this spatial signal pattern shaping is to simultaneously place a beam maximum toward the *signal-of-interest* (SOI) and ideally nulls toward directions of interfering signals or *signals-not-of-interest* (SNOIs). This process continuously changes to accommodate the incoming SOIs and SNOIs. The signal processor of the array must automatically adjust, from the collected information, the *weight vector* $\mathbf{w} = [w_1, w_2, \ldots, w_N]^T$ which corresponds to the complex amplitude excitation along each antenna element. It is usually convenient to represent the signal envelopes and the applied weights in their complex envelope form [62].

This relationship is represented by

$$\mathbf{r}(t) = \mathbf{Re}\left[\mathbf{x}(t)e^{j\omega_c t}\right] \tag{6.2}$$

where ω_c is the angular frequency of operation and $\mathbf{x}(t)$ is the complex envelope of the received real signal $\mathbf{r}(t)$. The incoming signal is weighted by the array pattern and the output is represented by

$$y(t) = \mathbf{Re}\left[\sum_{n=1}^{N} w_n^*(t)x_n(t)e^{j\omega_c t}\right] = \mathbf{Re}\left[\mathbf{w}^H(t)\mathbf{x}(t)e^{j\omega_c t}\right] \tag{6.3}$$

where n indicates each of the array elements and $\mathbf{w}^H(t)\mathbf{x}(t)$ is the complex envelope representation of $y(t)$. Since for any modern electronic system, signal processing is performed in discrete-time, the weight vector $\mathbf{w}$ combines linearly the collected discrete samples to form a single signal output expressed as

$$y(k) = \sum_{n=1}^{N} w_n^* x_n(k) = \mathbf{w}^H \mathbf{x}(k) \tag{6.4}$$

where k denotes discrete time index of the received signal sample being considered. The concept of beamforming is applicable in both continuous-time and discrete-time signals. Therefore, each element of the receiving antenna array possesses the necessary electronics to downconvert the received signal to baseband and for *analog-to-digital* (AD) conversion for *digital beamforming*. To simplify the analysis of this chapter, only baseband equivalent complex signal envelopes along with discrete-time processing will be considered herein.

Various adaptive algorithms have already been developed to calculate the optimal weight coefficients that satisfy several criteria or constraints. Once the beamforming weight vector $\mathbf{w}$ is calculated, the response of this spatial filter is represented by the *antenna radiation pattern* (*beampattern*) for all directions, which is expressed as

$$P(\theta) = \left|\mathbf{w}^H(\theta)\mathbf{a}(\theta)\right|^2. \tag{6.5}$$

In (6.5), $P(\theta)$ represents the average power of the spatial filter output when a single, unity-power signal arrives from angle θ [134]. With proper control of the magnitude and phase in $\mathbf{w}$, the pattern will exhibit a main beam in the direction of the desired signal and, ideally, nulls toward the direction of the interfering signals.

6.1   THE CLASSICAL BEAMFORMER

In *classical beamforming*, the beamforming weight is set to be equal to the array response vector of the desired signal. For any particular direction θ_0, the antenna pattern formed using the weight vector $\mathbf{w}_b = \mathbf{a}(\theta_0)$ has the maximum gain in this direction compared to any other possible weight vector of the same magnitude. This is accomplished because $\mathbf{w}_b$ adjusts the phases of the incoming signals arriving at each antenna element from a given direction θ_0 so that they add in-phase (or constructively). Because all the elements of the beamforming weight vector are basically phase shifts with unity magnitude, the system is commonly referred to as *phased array*. Mathematically, the desired response of the method can be justified by the Cauchy–Schwartz inequality

$$\left|\mathbf{w}^H(\theta)\mathbf{a}(\theta_0)\right|^2 \leq \|\mathbf{w}\|^2\|\mathbf{a}(\theta_0)\|^2 \tag{6.6}$$

for all vectors $\mathbf{w}$, with equality holding if and only if $\mathbf{w}$ is proportional to $\mathbf{a}(\theta_0)$ [134]. In the absence of array ambiguity, the effective pattern in (6.5) possesses a global maximum at θ_0. Even though the classical beamformer is the ideal choice to direct the maximum of the beampattern toward the direction of a SOI, since the complex weight vector $\mathbf{w}$ can be easily derived in closed form, it lacks the additional ability to place nulls toward any present SNOIs, often required in pragmatic scenarios [59]. This is obvious when observing the expression in (6.5) where, besides the look direction θ_0, control of the beampattern cannot be achieved in the rest of the angular region of interest. Thus, to accommodate all the requirements, a more advanced spatial processing technique is necessary to be applied.

To demonstrate this principle, we consider a six-element uniform linear array of omni-directional elements with half-wavelength spacing between adjacent elements. We assume that three equal-power uncorrelated sources are transmitting signals toward the array. Furthermore, the SOI is in the $\theta = 30°$ direction, toward which it is desired for the beampattern to possess its maximum and ideally also two nulls (for the two SNOIs) toward $\theta = -45°$ and $\theta = 0°$. Fig. 6.1 shows the two beamformed patterns: one using the classical beamformer [59] and the other based on a specific adaptive beamforming algorithm. As expected, the classical beam-former directs its maximum toward the direction of the SOI but fails to form nulls toward the directions of the SNOIs, since it does not have control of the beampattern beyond θ_0, whereas the adaptive beamforming algorithms achieve simultaneously to form a maximum toward the direction of the SOI and place nulls in the directions of the SNOIs.

6.2   STATISTICALLY OPTIMUM BEAMFORMING WEIGHT VECTORS

Depending on how the beamforming weights are chosen, beamformers can be classified as data independent or statistically optimum. The weights in a data independent beamformer do not depend on the received array data and are chosen to present a specified response for all signal and interference scenarios [22]. In practice, propagating waves are perturbed by the propagating medium or the receive mechanism. In this case, the plane wave assumption may no longer hold

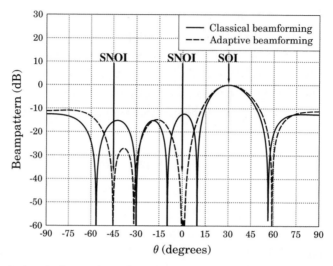

FIGURE 6.1: Classical and adaptive beamforming.

and weight vectors based on plane-wave delays between adjacent elements will not combine coherently the waves of the desired signal [22].

Matching of a randomly perturbed signal with arbitrary characteristics can be realized only in a statistical sense by using a matrix weighting of input data which adapts to the received signal characteristics [62]. This is referred to as *statistically optimum beamforming*. In this case, the weight vectors are chosen based on the statistics of the received data. The weights are selected to optimize the beamformer response so that the array output contains minimal contributions due to noise and signals arriving from directions other than that of the desired signal [144].

Any possible performance degradation may result due to a deviation of the actual operating conditions from the assumed ideal and can be minimized by the use of complementary methods that introduce constraints [22]. Due to the interest in applying array signal processing techniques in cellular communications, where mobile units can be located anywhere in the cell, statistically optimum beamformers provide the ability to adapt to the statistics of different subscribers. There exist different criteria for determining statistically-optimum beamformer weights, several of which are reviewed in this chapter.

6.2.1   The Maximum SNR Beamformer

The maximum SNR beamformer is essentially an extension of the classical beamformer. In the presence of noise, the weight vector $\mathbf{w}$ that maximizes the *Signal to Noise Ratio* (SNR) is given by [19]

$$\mathbf{w}_{maxSNR} = \mathbf{R}_{nn}^{-1}\mathbf{a}(\theta_0) \tag{6.7}$$

where $\mathbf{R}_{nn}$ is the noise covariance matrix. This beamforming weight vector gives an output with the maximum SNR when the noise covariance matrix is known. When the noise is spatially white, i.e., the noise covariance is a multiple of the identity matrix $\mathbf{I}$, the maximum SNR beamformer is equivalent to the classical beamformer [19]. Since only the desired signal direction is taken into account when calculating the beamformer weight vector, as in the case of the classical beamformer, the maximum SNR beamformer works adequately in a single-source scenario but cannot deal satisfactorily with interfering sources [19].

6.2.2   The Multiple Sidelobe Canceller and the Maximum SINR Beamformer

In the case of more than one user in the communication system, it is often desired to suppress the interfering signals, in addition to noise, using appropriate signal processing techniques. There are some intuitive methods to accomplish this, for example, the *multiple sidelobe canceller* (MSC) [144]. The basic idea of the MSC is that the conventional beamforming weight vectors for each of the signal sources are first calculated and the final beamforming vector is a linear combination of them in a way that the desired signal is preserved whereas all the interference components are eliminated [19]. The method for a particular geometry (ULA) has been already analyzed in a previous chapter to demonstrate the functional principle of smart antennas. MSC has some limitations, however. For instance, for a large number of interfering signals it cannot cancel all of them adequately and can result in significant gain for the noise component [144]. The solution to these limitations is the maximum SINR beamformer that maximizes the output signal to interference and noise power ratio.

Recall that the output of the beamformer is given by [19]

$$y = \mathbf{w}^H \mathbf{x} = \mathbf{w}^H (\mathbf{s} + \mathbf{i} + \mathbf{n}) = y_s + y_{IN} \qquad (6.8)$$

where all the components collected by the array at a single observation instant are $N \times 1$ complex vectors and are classified as: $\mathbf{s}$ is the desired signal component arriving from DOA θ_0, $\mathbf{i} = \sum_{i=1}^{I} \mathbf{s}_i$ is the interference component (assuming I such sources to be present), and $\mathbf{n}$ is the noise component. In (6.8), we also separate the desired signal array response weighted output, $y_s = \mathbf{w}^H \mathbf{s}$, and the interference-plus-noise total array response, $y_{IN} = \mathbf{w}^H (\mathbf{i} + \mathbf{n})$. Consequently, the weighted array signal output power is [22]

$$\mathcal{E}\left\{|y_s|^2\right\} = \mathbf{w}^H \mathcal{E}\left\{\mathbf{s}\mathbf{s}^H\right\} \mathbf{w} = \mathbf{w}^H \mathbf{R}_{ss} \mathbf{w} \qquad (6.9)$$

where $\mathbf{R}_{ss}$ is the autocovariance matrix of the signal vectors $\mathbf{s}$ and the weighted interference-plus-noise output power is [22]

$$\mathcal{E}\left\{|y_{IN}|^2\right\} = \mathbf{w}^H \mathcal{E}\left\{|\mathbf{i} + \mathbf{n}|^2\right\} \mathbf{w} = \mathbf{w}^H \mathbf{R}_{IN} \mathbf{w} \qquad (6.10)$$

where $\mathbf{R}_{IN}$ is the autocovariance matrix of the vectors $\mathbf{n} + \mathbf{i}$. Therefore, the weighted output SINR can be expressed as [22]

$$SINR = \frac{\mathcal{E}\{|y_s|^2\}}{\mathcal{E}\{|y_{IN}|^2\}} = \frac{\mathbf{w}^H \mathbf{R}_{ss} \mathbf{w}}{\mathbf{w}^H \mathbf{R}_{IN} \mathbf{w}}. \qquad (6.11)$$

With appropriate factorization of $\mathbf{R}_{IN}$ and manipulation of the SINR expression, the maximization problem can be recognized as an eigen-decomposition problem. The expression for $\mathbf{w}$ that maximizes the SINR is found to be [22]

$$\mathbf{w}_{maxSINR} = \mathbf{R}_{IN}^{-1} \mathbf{a}(\theta_0). \qquad (6.12)$$

This is the statistical optimum solution in maximizing the output SINR in an interference plus noise environment, but it requires a computationally intensive inversion of $\mathbf{R}_{IN}$, which may be problematic when the number of elements in the antenna array is large [19].

6.2.3 Minimum Mean Square Error (MMSE)

If sufficient knowledge of the desired signal is available, a reference signal d can then be generated. These reference signals are used to determine the optimal weight vector $\mathbf{w}_{MSE} = [w_1, w_2, \ldots, w_N]^T$. This is done by *minimizing* the *mean square error* of the reference signals and the outputs of the N-element antenna array [145]. The concept of reference signal use in adaptive antenna system was first introduced by Widrow in [145] where he described several pilot-signal generation techniques. One of the proposed techniques used a two-mode adaptation process whereby the transmitter alternated between sending a known pilot signal and actual data. The receiver had knowledge of the pilot signal and used it as the desired response for the LMS adaptive algorithm (described later in this chapter). During actual data transmission, adaptation would be switched off and the weights would coast until the pilot signal was turned back on. While an adaptive antenna utilizing this technique was probably never constructed, the concept provided the necessary impetus which eventually grew into actual hardware implementations [146].

For beamforming considerations, the reference signal is usually obtained by a periodic transmission of a training sequence, which is *a priori* known at the receiver and is referred to as *temporal reference*. Note that information about the direction of the signal of interest is usually referred to as *spatial reference*. The temporal reference is of vital importance in a fading environment due to lack of angle of arrival information [70]. As described by Compton [147], the adaptive array reference signal need not necessarily be an exact replica of the desired signal, even though this is what occurs in most of the cases. In general, it can be unknown but needs to be correlated with the desired signal and uncorrelated with any possible interference. Compton goes on to describe several experimental antenna systems designed for use with spread spectrum

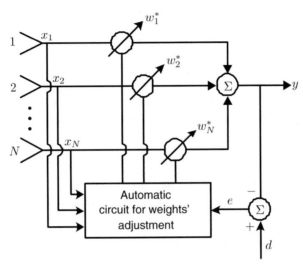

FIGURE 6.2: Reference signal adaptive antenna [22].

signals where the spreading sequence provides the necessary discrimination between desired signal and interference. A tutorial discussion on adaptive beamformers with self-generated reference signals can be found in [146].

A block diagram of an adaptive system using reference signals is shown in Fig. 6.2. At each observation instance k, the error $e(k)$ between the reference signal $d(k)$ and the weighted array output $y(k)$ is given by

$$e(k) = d(k) - y(k) = d(k) - \mathbf{w}^H\mathbf{x}(k). \tag{6.13}$$

Mathematically, the MMSE criterion can be expressed as

$$\min_{\mathbf{w}} \mathcal{E}\left\{J_{\mathbf{w},\mathbf{w}^*}\right\}$$

where $J_{\mathbf{w},\mathbf{w}^*} = |e(k)|^2$ denotes the real-valued objective function of the weight vector $\mathbf{w}$ to be solved ($\mathbf{w}^*$ is the conjugate of $\mathbf{w}$). The maximum rate of change of $J_{\mathbf{w},\mathbf{w}^*}$ is given by $\frac{\partial J_{\mathbf{w},\mathbf{w}^*}}{\partial \mathbf{w}^*}$ [83, 148]. In order to get a meaningful result, the objective function needs to have explicit dependency on the conjugate of the weight vector [23]. Usually this simply translates into changing transposition to conjugate transposition (or Hermitian). For a more detailed discussion on the topic, see [83, 148]. Therefore, we have

$$\frac{\partial J_{\mathbf{w},\mathbf{w}^*}}{\partial \mathbf{w}^*} = \frac{\partial \left\{\left[d(k) - \mathbf{w}^H\mathbf{x}(k)\right]^H \left[d(k) - \mathbf{w}^H\mathbf{x}(k)\right]\right\}}{\partial \mathbf{w}^*} \tag{6.14}$$

$$= -2e^*(k)\mathbf{x}(k).$$

To minimize the objective function, we set (6.14) to zero. Considering additionally the expectation value of the minimum of $J_{\mathbf{w},\mathbf{w}^*}$, it yields

$$2\mathbf{R}_{xx}\mathbf{w} - 2\mathbf{r}_{xd} = 0 \qquad (6.15)$$

where $\mathbf{R}_{xx} = \mathcal{E}\left\{\mathbf{x}\mathbf{x}^H\right\}$ is the signal autocovariance matrix and $\mathbf{r}_{xd} = \mathcal{E}\left\{\mathbf{x}d^*\right\}$ is the reference signal covariance vector. Thus, the optimal MMSE weight solution is given by

$$\mathbf{w}_{\text{MSE}} = \mathbf{R}_{xx}^{-1}\mathbf{r}_{xd}. \qquad (6.16)$$

and is usually referred to as the Wiener–Hopf solution. One disadvantage using this method is the generation of an accurate reference signal based on limited knowledge at the receiver [22].

6.2.4   Direct Matrix Inversion (DMI)

If the desired and interference signals are known *a priori*, (6.16) provides the most direct and fastest solution to compute the optimal weights. However, the signals are not known exactly since the signal environment undergoes frequent changes. Thus, the signal processing unit must continually update the weight vector to meet the new requirements imposed by the varying conditions [98]. This need to update the weight vector, without *a priori* information, leads to estimating the covariance matrix, $\mathbf{R}_{xx}$, and the cross-correlation vector, $\mathbf{r}_{xd}$, in a finite observation interval. Note that this is a block-adaptive approach where the statistics are estimated using temporal blocks of the array data [70]. The adaptivity is achieved via a sliding window, say of length L symbols. The estimates $\hat{\mathbf{R}}_{xx}$ and $\hat{\mathbf{r}}_{xd}$ can be evaluated as:

$$\hat{\mathbf{R}}_{xx} = \frac{1}{L}\sum_{i=N_1}^{N_2}\mathbf{x}(i)\mathbf{x}^H(i) \qquad (6.17a)$$

$$\hat{\mathbf{r}}_{xd} = \frac{1}{L}\sum_{i=N_1}^{N_2}\mathbf{x}(i)d^*(i) \qquad (6.17b)$$

where N_1 and N_2 are, respectively, the lower and upper limits of the observation interval such that $N_2 = N_1 + L - 1$. Thus, the estimate for the weight vector is given by

$$\hat{\mathbf{w}}_{\text{MSE}} = \hat{\mathbf{R}}_{xx}^{-1}\hat{\mathbf{r}}_{xd}. \qquad (6.18)$$

The advantage of the method is that it converges faster than any adaptive method, and the rate of convergence does not depend on the power level of the signals. However, two major problems are associated with the matrix inversion. First, the increased computational complexity cannot be easily overcomed through the use of integrated circuits, and second, the use of finite-precision arithmetic and the necessity of inverting a large matrix may result in numerical instability.

6.2.5   Linearly Constrained Minimum Variance (LCMV)

In the MMSE criterion, the Wiener filter minimizes the MSE with no constraints imposed on the solution (i.e., the weights). However, it may be desirable, or even mandatory, to design a filter that minimizes a mean square criterion subject to a specific constraint. The LCMV constrains the response of the beamformer so that signals from the direction of interest are passed through the array with a specific gain and phase [149]. However it requires knowledge, or prior estimation, of the desired signal array response $\mathbf{a}(\theta_0)$ with DOA θ_0. Its weights are chosen to minimize the expected value of the output power/variance subject to the response constraints. That is [22]

$$\min_{\mathbf{w}} \left\{ \mathbf{w}^H \mathbf{R}_{xx} \mathbf{w} \right\} \quad \text{subject to} \quad \mathbf{C}^H \mathbf{w} = \mathbf{g}^*$$

where $\mathbf{C} \in \mathbb{C}^{N \times K}$ has K linearly independent constraints and $\mathbf{g} \in \mathbb{C}^{K \times 1}$ is the constraint response vector.

The constraints have an effect of preserving the desired signal while minimizing contributions to the array output due to interfering signals and noise arriving from directions other than that of interest [22]. The solution to this constrained optimization problem requires the use of the *Lagrange multiplier* vector $\mathbf{b} \in \mathbb{C}^K$. Letting $F(\mathbf{w}) = \mathbf{w}^H \mathbf{R}_{xx} \mathbf{w}$ be the cost function and $\mathbf{G}(w) = \mathbf{C}^H \mathbf{w} - \mathbf{g}^*$ be the constraint function, the following expression is formed [22]:

$$\begin{aligned} H(\mathbf{w}) &= \frac{1}{2} F(\mathbf{w}) + \mathbf{b}^H \mathbf{G}(w) \\ &= \frac{1}{2} \mathbf{w}^H \mathbf{R}_{xx} \mathbf{w} + \mathbf{b}^H \left(\mathbf{C}^H \mathbf{w} - \mathbf{g}^* \right). \end{aligned} \tag{6.19}$$

$F(\mathbf{w})$ has its minimum value at a point $\mathbf{w}$ subject to the constraint $\mathbf{G}(w) = \mathbf{C}^H \mathbf{w} - \mathbf{g}^* = 0$, i.e., when $H(\mathbf{w})$ is minimum. Therefore, to find the minimum point in equation (6.19), we differentiate with respect to $\mathbf{w}$ and set it equal to zero, which yields [22]:

$$\mathbf{w}_{\text{opt}} = -\mathbf{R}_{xx}^{-1} \mathbf{C} \mathbf{b}. \tag{6.20}$$

Substituting $\mathbf{w}_{\text{opt}}$ back into the constraint equation yields [22]

$$\mathbf{b} = -\left[\mathbf{C}^H \mathbf{R}_{xx}^{-1} \mathbf{C} \right]^{-1} \mathbf{g}^* \tag{6.21}$$

where the existence of $\left[\mathbf{C}^H \mathbf{R}_{xx}^{-1} \mathbf{C} \right]$ follows from the fact that $\mathbf{R}_{xx}$ is positive definite and $\mathbf{C}$ is full-rank. Therefore, the LCMV estimate of the weight vector is [22]

$$\mathbf{w}_{\text{opt}} = \mathbf{R}_{xx}^{-1} \left[\mathbf{C}^H \mathbf{R}_{xx}^{-1} \mathbf{C} \right]^{-1} \mathbf{g}^*. \tag{6.22}$$

As a special case, a requirement would be to force the beam pattern to be constant in the boresight direction; concisely, this can be stated mathematically as [150]

$$\min_{\mathbf{w}} \left\{ \mathbf{w}^H \mathbf{R}_{xx} \mathbf{w} \right\} \quad \text{subject to} \quad \mathbf{w}^H \mathbf{a}(\theta_0) = g^*$$

where g is a complex scalar which constrains the output response to $\mathbf{a}(\theta_0)$. In this case, the LCMV weight estimate is [22]

$$\mathbf{w}_{\text{opt}} = g^* \frac{\mathbf{R}_{xx}^{-1} \mathbf{a}(\theta_0)}{\mathbf{a}^H(\theta_0) \mathbf{R}_{xx}^{-1} \mathbf{a}(\theta_0)}. \qquad (6.23)$$

For the special case when $g = 1$ (i.e., the gain constant is unity), the optimum solution of (6.23) is termed as the *minimum variance distortionless response* (MVDR) beamformer, and it is also referred to as the *maximum likelihood method* (MLM) because the algorithm maximizes the likelihood function of the input signal [98].

The advantage of using LCMV criteria is its general constraint approach that permits extensive control over the adapted response of the beamformer [22]. It is a flexible technique that does not require knowledge of the desired signal autocovariance matrix $\mathbf{R}_{xx}$, the interference-plus-noise autocovariance matrix $\mathbf{R}_{IN}$, or any reference signal $d(k)$ [22]. A certain level of beamforming performance can be attained through the design of the beamformer, allowed by the constraint matrix [22]. However, the disadvantage of using LCMV criteria is the computation complexity of the constraint weight vector. There are several constraint designs for the LCMV performance such as point constraints, eigenvector constraints, etc., which are beyond the scope of the present discussion.

6.3 ADAPTIVE ALGORITHMS FOR BEAMFORMING

As previously shown, statistically optimum weight vectors for adaptive beamforming can be calculated by the Wiener solution. However, knowledge of the asymptotic second-order statistics of the signal and the interference-plus-noise was assumed. These statistics are usually not known but with the assumption of ergodicity, where the time average equals the ensemble average, they can be estimated from the available data [22]. For time-varying signal environments, such as wireless cellular communication systems, statistics change with time as the target mobile and interferers move around the cell. For the time-varying signal propagation environment, a recursive update of the weight vector is needed to track a moving mobile so that the spatial filtering beam will adaptively steer to the target mobile's time-varying DOA, thus resulting in optimal transmission/reception of the desired signal [22]. To solve the problem of time-varying statistics, weight vectors are typically determined by adaptive algorithms which adapt to the changing environment.

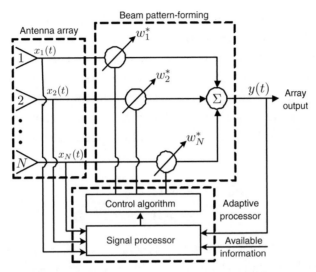

FIGURE 6.3: Functional diagram of an *N*-element adaptive array [22].

Fig. 6.3 shows a generic adaptive antenna array system consisting of an *N*-element antenna array with a real time adaptive array signal processor containing an update control algorithm. The data samples collected by the antenna array are fed into the signal processing unit which computes the weight vector according to a specific control algorithm.

Steady-state and transient-state are the two classifications of the requirement of an adaptive antenna array. These two classifications depend on whether the array weights have reached their steady-state values in a stationary environment or are being adjusted in response to alterations in the signal environment. If we consider that the reference signal for the adaptive algorithm is obtained by temporal reference, *a priori* known at the receiver during the actual data transmission, we can either continue to update the weights adaptively via a decision directed feedback or use those obtained at the end of the training period [70]. Several adaptive algorithms can be used such that the weight vector adapts to the time-varying environment at each sample; some of them are now reviewed. The text and tables, appearing in the descriptions of the adaptive algorithms 1–2 and 4–5 that follow, are in great part reproduced and adopted from [23] (pp. 9–15)[1].

6.3.1   The Least Mean-Square (LMS) Algorithm

The LMS algorithm [150, 151] is probably the most widely used adaptive filtering algorithm, being employed in several communication systems. It has gained popularity due to

---

[1]The material was reproduced with the courtesy and permission of the author of [23] who retains its copyright.

its low computational complexity and proven robustness [23]. It incorporates new obser-vations and iteratively minimizes linearly the mean-square error [62, 83, 145]. The LMS algorithm changes the weight vector $\mathbf{w}$ along the direction of the estimated gradient based on the negative steepest descent method [152]. By the quadratic characteristics of the mean square-error function $\mathcal{E}\{|e(k)|^2\}$ that has only one minimum, the steepest descent is guaranteed to converge. At adaptation index k, given a *mean-square-error* (MSE) function $\mathcal{E}\{|e(k)|^2\} = \mathcal{E}\{|d(k) - \mathbf{w}^H\mathbf{x}(k)|^2\}$, the LMS algorithm updates the weight vector according to [22]

$$\begin{aligned}\mathbf{w}(k+1) &= \mathbf{w}(k) - \frac{\mu}{2}\frac{\partial J_{\mathbf{w},\mathbf{w}^*}}{\partial \mathbf{w}^*}\\ &= \mathbf{w}(k) + \mu e^*(k)\mathbf{x}(k)\end{aligned} \tag{6.24}$$

where the rate of change of the objective function $J_{\mathbf{w},\mathbf{w}^*} = |e(k)|^2$ has been derived earlier in (6.14) and μ is a scalar constant which controls the rate of convergence and stability of the algorithm. In order to guarantee stability in the mean-squared sense, the step size μ should be restricted in the interval [22]

$$0 < \mu < \frac{2}{\lambda_{\max}} \tag{6.25}$$

where $\lambda_{\max}$ is the maximum eigenvalue of $\mathbf{R}_{xx}$. Alternatively, in terms of the total power of the vector $\mathbf{x}$ [22]

$$\lambda_{\max} \leq \text{trace}\{\mathbf{R}_{xx}\} \tag{6.26}$$

where $\text{trace}\{\mathbf{R}_{xx}\} = \sum_{i=1}^{N}\mathcal{E}\{x_i^2\}$ is the total input power. Therefore, a condition for satisfactory Wiener solution convergence of the mean of the LMS weight vector is [22]

$$0 < \mu < \frac{2}{\sum_{i=1}^{N}\mathcal{E}\{x_i^2\}} \tag{6.27}$$

where N is the number of elements in the array. The pseudo-code for the LMS algorithm is shown in Table 6.1 [23]. A normalized version of the LMS algorithm, the NLMS algorithm [150, 153, 154], also referred to as the *projection algorithm* (PA) in the control literature [155], is obtained by substituting the step size in (6.24) with the time-varying step size $\mu/\|\mathbf{x}(k)\|^2$, where $0 < \mu < 2$ [154]. A significant drawback from the use of the LMS and the NLMS algorithms is their slow convergence for colored noise input signals [23].

The LMS algorithm is a member of a family of stochastic gradient algorithms since the instantaneous estimate of the gradient vector is a random vector that depends on the input data

TABLE 6.1: The Least Mean-Square Algorithm [22]

LMS ALGORITHM

for each k
{

$$e(k) = d(k) - \mathbf{w}^H(k)\mathbf{x}(k)$$
$$\mathbf{w}(k+1) = \mathbf{w}(k) + \mu e^*(k)\mathbf{x}(k)$$

}

vector $\mathbf{x}(k)$ [156]. It requires about $2N$ complex multiplications per iteration, where N is the number of weights (elements) used in the adaptive array. The convergence characteristics of the LMS depend directly on the eigenstructure of $\mathbf{R}_{xx}$ [22]. Its convergence can be slow if the eigenvalues are widely spread. When the covariance matrix eigenvalues differ substantially, the algorithm convergence time can be exceedingly long and highly data dependent [62]. Therefore, depending on the eigenvalue spread, the LMS algorithm may not have sufficient iteration time for the weight vector to converge to the statistically optimum solution and adaptation in real time to the time-varying environment will not be able to be performed [22]. In addition, employing the LMS algorithm, it is assumed that sufficient knowledge of the desired signal is known so as to generate reference signal sequences. However, acquiring this knowledge could be very expensive for wireless communication systems, especially in fast-fading scenarios [22]. In cases where the convergence speed of the LMS algorithm is not satisfied, the following algorithms may serve as acceptable alternatives.

6.3.2 The Recursive Least-Squares (RLS) Algorithm

Unlike the LMS algorithm [150, 157] which uses the method of steepest descent to update the weight vector, the RLS adaptive algorithm approximates the Wiener solution directly using the *method of least* squares to adjust the weight vector, without imposing the additional burden of approximating an optimization procedure [144]. In the method of least squares, the weight vector $\mathbf{w}(k)$ is chosen so as to minimize a cost function that consists of the sum of error squares over a time window, i.e., the *least-square* (LS) solution is minimized recursively [23]. In the method of steepest-descent, on the other hand, the weight vector is chosen to minimize the ensemble average of the error squares. The recursions for the most common version of the RLS algorithm, which is presented in its standard form in Table 6.2 [23], are a result of the *weighted*

TABLE 6.2: The Recursive Least-Squares Algorithm [23]

RLS ALGORITHM
$\mathbf{R}^{-1}(0) = \delta^{-1}\mathbf{I}$, δ small positive constant and $\mathbf{I}$ the $N \times N$ identity matrix for each k { $\qquad \mathbf{k}(k) = \mathbf{R}^{-1}(k-1)\mathbf{x}(k)$ $\qquad \kappa(k) = \frac{\mathbf{k}(k)}{\lambda + \mathbf{x}^H(k)\mathbf{k}(k)}$ $\qquad \mathbf{R}^{-1}(k) = \frac{1}{\lambda}\left[\mathbf{R}^{-1}(k-1) - \frac{\mathbf{k}(k)\mathbf{k}^H(k)}{\lambda + \mathbf{x}^H(k)\mathbf{k}(k)}\right]$ $\qquad e(k) = d(k) - \mathbf{w}^H(k)\mathbf{x}(k)$ $\qquad \mathbf{w}(k+1) = \mathbf{w}(k) + e^*(k)\kappa(k)$ }

least-squares (WLS) objective function

$$J_{\mathbf{w},\mathbf{w}^*} = \sum_{i=1}^{k} \lambda^{k-1}|e(i)|^2 \qquad (6.28)$$

where the error signal $e(i)$ has been defined earlier and $0 < \lambda \leq 1$ is an exponential scaling factor which determines how quickly the previous data are de-emphasized [156] and is referred to as the *forgetting factor* [23]. Usually, λ is chosen close to, but less than, unity. However, in a stationary environment λ should be equal to 1, since all data past and present should have equal weight [156]. Differentiating the objective function $J_{\mathbf{w},\mathbf{w}^*}$ with respect to $\mathbf{w}^*$ and solving for the minimum yields [23]

$$\left[\sum_{i=1}^{k} \lambda^{k-1}\mathbf{x}(i)\mathbf{x}^H(i)\right]\mathbf{w}(k) = \sum_{i=1}^{k} \lambda^{k-1}\mathbf{x}(i)d^*(i). \qquad (6.29)$$

Furthermore, defining the quantities [23]

$$\mathbf{R}(k) = \sum_{i=1}^{k} \lambda^{k-1}\mathbf{x}(i)\mathbf{x}^H(i) \qquad (6.30)$$

and

$$\mathbf{p}(k) = \sum_{i=1}^{k} \lambda^{k-1}\mathbf{x}(i)d^*(i) \qquad (6.31)$$

the solution is obtained as [23]

$$\mathbf{w}(k) = \mathbf{R}^{-1}(k)\mathbf{p}(k). \tag{6.32}$$

The recursive implementations are a result of the formulations

$$\mathbf{R}(k) = \lambda \mathbf{R}(k-1) + \mathbf{x}(k)\mathbf{x}^H(k) \tag{6.33}$$

and

$$\mathbf{p}(k) = \lambda \mathbf{p}(k-1) + \mathbf{x}(k)d^*(k). \tag{6.34}$$

The inverse $\mathbf{R}^{-1}(k)$ can be obtained recursively in terms of $\mathbf{R}^{-1}(k-1)$ using the *matrix inversion lemma*[2] [151], thus avoiding direct inversion of $\mathbf{R}(k)$ at each time instant k.

An important feature of the RLS algorithm is that it utilizes information contained in the input data, extending back to the time instance the algorithm was initiated. The resulting rate of convergence is therefore typically an order of magnitude faster than the simple LMS algorithm. This improvement in performance, however, is achieved at the expense of a large increase in computational complexity. The RLS algorithm requires $4N^2 + 4N + 2$ complex multiplications per iteration, where N is the number of weights used in the adaptive array. Other drawbacks associated with its implementation are potential divergence behavior in a finite-precision environment and stability problems that usually result in loss of symmetry and positive definiteness of the matrix $\mathbf{R}^{-1}(k)$ [23].

6.3.3 The Constant-Modulus (CM) Algorithm

Many communication signals, frequency or phase modulated, such as FM, CPFSK modulation, and square pulse-shaped complex pulse amplitude modulation (PAM) have a constant complex envelope [159]. This property is usually referred to as the *constant modulus* (CM) signal property. For these types of communication signals, one can take advantage of the prior knowledge of this characteristic and specify the adaptation algorithm to achieve a desired steady state response from the array [160]. The *constant-modulus algorithm* is the most well-known algorithm of this kind. It is suitable for the transmission of a modulated signal over the wireless channel, since noise and interference corrupt the CM property of the desired signal [159]. A signal traveling through a frequency selective channel is almost sure to also lose its constant modulus property.

---

[2]If $\mathbf{A}$, $\mathbf{B}$, $\mathbf{C}$, and $\mathbf{D}$ are matrices with dimensions $n \times n$, $n \times m$, $m \times m$, and $m \times n$, respectively, then $[\mathbf{A} + \mathbf{BCD}]^{-1} = \mathbf{A}^{-1} - \mathbf{A}^{-1}\mathbf{B}\left[\mathbf{DA}^{-1}\mathbf{B} + \mathbf{C}^{-1}\right]^{-1}\mathbf{DA}^{-1}$, provided that the inverses of the indicated square matrices exist. For a proof of the lemma, the reader is referred to [152]. A special case, known as *Woodbury's identity*, results for $\mathbf{B}$ being an $n \times 1$ column vector $\mathbf{u}$, $\mathbf{C}$ a scalar of unity and $\mathbf{D}$ a $1 \times n$ row vector $\mathbf{u}^T$. Then $[\mathbf{A} + \mathbf{uu}^T]^{-1} = \mathbf{A}^{-1} - \frac{\mathbf{A}^{-1}\mathbf{uu}^T\mathbf{A}^{-1}}{1+\mathbf{u}^T\mathbf{A}^{-1}\mathbf{u}}$ [158].

Thus, the CM provides an indirect measure of the quality of the filtered signal. It adjusts the weight vector of the adaptive array so as to minimize the variation of the desired signal at the array. After the algorithm converges, a beam is steered in the direction of the signal of interests, whereas nulls are placed in the direction of interference. In general, the CM algorithm seeks a beamformer weight vector that minimizes a cost function of the form

$$J_{p,q} = \mathcal{E}\left\{||y(k)|^p - 1|^q\right\}. \tag{6.35}$$

Equation (6.35) describes a family of cost functions. The convergence of the algorithm depends on the coefficients p and q in (6.35). A particular choice of p and q yields a specific cost function called the (p,q) *CM cost function*. The $(1, 2)$ and $(2, 2)$ CM cost functions are the most popular. The objective of CM beamforming is to restore the array output $y(k)$ to a constant envelope signal. Using the method of steepest descent, the weight vector is updated using the following recursive equation,

$$\mathbf{w}(k+1) = \mathbf{w}(k) - \mu\nabla_{\mathbf{w},\mathbf{w}^*}\left(J_{p,q}\right) \tag{6.36}$$

where the step-size parameter has been denoted by μ. When the $(1,2)$ CM function is used, the gradient vector is given by [156]

$$\nabla_{\mathbf{w},\mathbf{w}^*}(J_{1,2}) = \frac{\partial J_{1,2}}{\partial \mathbf{w}^*} = \mathcal{E}\left[\mathbf{x}(k)\left(y(k) - \frac{y(k)}{|y(k)|}\right)^*\right]. \tag{6.37}$$

Ignoring the expectation operation in (6.37), the instantaneous estimate of the gradient vector can be written as

$$\nabla_{\mathbf{w},\mathbf{w}^*}\left(\hat{J}_{1,2}(k)\right) = \mathbf{x}(k)\left[y(k) - \frac{y(k)}{|y(k)|}\right]^* \tag{6.38}$$

and therefore, using (6.38), the resulting weight vector is given by

$$\begin{aligned}
\mathbf{w}(k+1) &= \mathbf{w}(k) - \mu\left[y(k) - \frac{y(k)}{|y(k)|}\right]^*\mathbf{x}(k) = \\
&= \mathbf{w}(k) + \mu e^*(k)\mathbf{x}(k)
\end{aligned} \tag{6.39}$$

where $e(k) = y(k)/|y(k)| - y(k)$. Comparing the CM and the LMS algorithms, we notice that they are very similar to each other. The term $\frac{y(k)}{|y(k)|}$ in CM plays the same role as the desired signal $d(k)$ in the LMS. However, the reference signal $d(k)$ must be sent from the transmitter to the receiver and must be known for both the transmitter and receiver if the LMS algorithm is used. The CM algorithm does not require a reference signal to generate the error signal at the receiver [156]. Several other properties of the constant modulus algorithm are discussed in [161]. The pseudo-code for the CM (1,2) algorithm is shown in Table 6.3.

TABLE 6.3: The Constant-Modulus Algorithm [23]
(1,2) CM ALGORITHM

for each k
{
$\qquad y(k) = \mathbf{w}^H(k)\mathbf{x}(k)$
$\qquad e(k) = \frac{y(k)}{\|y(k)\|} - y(k)$
$\qquad \mathbf{w}(k+1) = \mathbf{w}(k) + \mu e^*(k)\mathbf{x}(k)$
}

6.3.4   The Affine-Projection (AP) Algorithm

It is well known that the *normalized* LMS algorithm often converges faster than the basic LMS algorithm and in many times can effectively replace the RLS algorithm [23]. Examples of such low-complexity algorithms are the *binormalized data-reusing least mean-square* (BN-DRLMS) [162], the *normalized new data-reusing* (NNDR) [163], and the *affine-projection* (AP) [164–166] algorithms. Studies have shown that the idea of reutilizing past and present information in the coefficient update, referred to as data-reusing, to be a promising approach in achieving balance between convergence speed and computational complexity of the algorithm [23]. The BNDRLMS algorithm utilizes current and past data-pairs in its update. The relationships between a number of reusing algorithms are addressed in [167]. The AP projection algorithm can be seen as a general normalized data-reusing algorithm that reuses an arbitrary number of data-pairs. It updates its coefficient vector such that the new solution belongs to the intersection of P hyperplanes defined by the present and the $P-1$ previous data pairs $\{\mathbf{x}(i), d(i)\}_{i=k-P+1}^{k}$. The optimization criterion used for the derivation of the AP algorithm is given by

$$\mathbf{w}(k+1) = \arg\min_{\mathbf{w}} \|\mathbf{w} - \mathbf{w}(k)\|^2 \quad \text{subject to} \quad \mathbf{d}(k) = \mathbf{X}^T(k)\mathbf{w}^*$$

where

$$\mathbf{d}(k) = [d(k), d(k-1), \ldots, d(k-P+1)]^H \text{ and} \qquad (6.40a)$$
$$\mathbf{X}(k) = [\mathbf{x}(k), \mathbf{x}(k-1), \ldots, \mathbf{x}(k-P+1)]. \qquad (6.40b)$$

The updating equations for the AP algorithm obtained as the minimization problem in (6.3.4) are presented in Table 6.4 [23]. To control stability, convergence, and final error, a step size μ is introduced where $0 < \mu < 2$ [165]. To improve robustness, a diagonal matrix $\delta\mathbf{I}$ is used to

> **TABLE 6.4:** The Affine-Projection Algorithm [23]
>
> ### AP ALGORITHM
>
> for each k
> {
> $$e(k) = d(k) - \mathbf{X}^T(k)\mathbf{w}^*(k)$$
> $$t(k) = \left[\mathbf{X}^H(k)\mathbf{X}(k) + \delta\mathbf{I}\right]^{-1}\mathbf{e}^*(k)$$
> $$\mathbf{w}(k+1) = \mathbf{w}(k) + \mu\mathbf{X}(k)t(k)$$
> }

regularize the inverse matrix in the AP algorithm, where δ is a small positive constant and $\mathbf{I}$ is an $N \times N$ identity matrix [23].

6.3.5   The Quasi-Newton (QN) Algorithm

The fast convergence of the RLS algorithm relies on the estimation of the inverse of the correlation matrix $\mathbf{R}^{-1}(k)$ which is required to remain symmetric and positive definite for the algorithm's stability [23]. However, implementation in finite precision may cause $\mathbf{R}^{-1}(k)$ to become indefinite [168]. One algorithm that provides convergence speed comparable to that of the RLS algorithm, but is guaranteed to be stable even under high input-signal correlation and fixed-point short-wordlength arithmetic, is the quasi-Newton (QN) algorithm [168, 169]. In the QN algorithm, the weight vector is updated as

$$\mathbf{w}(k+1) = \mathbf{w}(k) + \mu(k)\mathbf{h}(k) \tag{6.41}$$

where $\mu(k)$ is a step size obtained through an exact line search, and $\mathbf{h}(k)$ is the direction of the update given by

$$\mathbf{h}(k) = -\mathbf{R}^{-1}(k-1)\frac{\partial J_{\mathbf{w},\mathbf{w}^*}}{\partial \mathbf{w}^*} \tag{6.42}$$

where the cost function is once more $J_{\mathbf{w},\mathbf{w}^*} = |e(k)|^2$. Performing an exact line search results in a step size [168]

$$\mu(k) = \frac{1}{2\mathbf{x}^H(k)\mathbf{R}^{-1}(k-1)\mathbf{x}(k)}. \tag{6.43}$$

The update of $\mathbf{R}^{-1}(k-1)$ is crucial for the numerical behavior of the QN algorithm, and different approximations lead to different QN algorithms [23]. For an approximation of $\mathbf{R}^{-1}(k-1)$,

TABLE 6.5: The Quasi-Newton Algorithm [23]

QN ALGORITHM

for each k

{

$\quad e(k) = d(k) - \mathbf{w}^H(k)\mathbf{x}(k)$

$\quad \mathbf{t}(k) = \mathbf{R}^{-1}(k-1)\mathbf{x}(k)$

$\quad \tau(k) = \mathbf{x}^H(k)\mathbf{t}(k)$

$\quad \mu(k) = \frac{1}{2\tau(k)}$

$\quad \mathbf{R}^{-1}(k) = \mathbf{R}^{-1}(k-1) + \frac{[\mu(k)-1]}{\tau(k)}\mathbf{t}(k)\mathbf{t}^H(k)$

$\quad \mathbf{w}(k+1) = \mathbf{w}(k) + \alpha\frac{e^*(k)}{\tau(k)}\mathbf{t}(k)$

}

which is robust and remains positive definite even for highly correlated input signals and short wordlength arithmetic, as given in [168], the QN algorithm can be implemented as shown in Table 6.5 [23]. In Table 6.5 a positive constant α is used to control the speed of convergence and the misadjustment. Convergence in the mean and the mean-squared sense of the weight vector is guaranteed for $0 < \alpha < 2$ provided that $\mathbf{R}^{-1}(k-1)$ is positive definite [168–170].

CHAPTER 7

Integration and Simulation of Smart Antennas

Unlike most of the work for smart antennas that covered each area individually (antenna-array design, signal processing, communications algorithms and network *throughput*), the work in this chapter may be considered as an effort on smart antennas that examines and integrates antenna array design, the development of signal processing algorithms (for angle of arrival estimation and adaptive beamforming), strategies for combating fading, and the impact on the network *throughput* [24, 171–174]. In particular, this work considers problems dealing with the impact of the antenna design on the network *throughput*. In addition, fading channels and tradeoffs between diversity combining and adaptive beamforming are examined as well as channel coding to improve the system performance.

7.1 OVERVIEW

The main goal of this chapter and reported in [24, 171–174], is to design smart antennas for Mobile Ad-Hoc Network (MANET) devices operating at a frequency of 20 GHz. This objective was instrumental in selecting elements that can conform to the geometry of the device and an array architecture that could control the radiation pattern both in the azimuth and elevation directions. Consequently, this led to the selection of microstrip patches arranged in a planar configuration. In addition, the number of radiating elements was chosen to meet beamwidth requirements while maintaining reasonable cost and complexity for hardware implementation.

To analyze the average network throughput, a channel access protocol was proposed for MANETs employing smart antennas. The proposed protocol was based on the MAC protocol of IEEE 802.11 WLANs for TDMA environment [175].

Results showed that network *throuput* was influenced by both the number of elements in a planar antenna array and different array designs (uniform, Tschebyscheff, adaptive). Moreover, the network throughput analysis was extended to impose guidelines on the beamforming algorithm convergence rate. Finally, the performance of the adaptive algorithms, i.e., the DMI

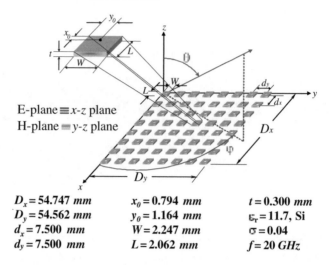

$D_x = 54.747$ mm	$x_0 = 0.794$ mm	$t = 0.300$ mm
$D_y = 54.562$ mm	$y_0 = 1.164$ mm	$\varepsilon_r = 11.7$, Si
$d_x = 7.500$ mm	$W = 2.247$ mm	$\sigma = 0.04$
$d_y = 7.500$ mm	$L = 2.062$ mm	$f = 20$ GHz

FIGURE 7.1: Planar-array configuration.

algorithm and the LMS algorithm, in Rayleigh-fading channels was examined. The material of this chapter is primarily derived from [24, 59, 171–174].

7.2   ANTENNA DESIGN

The type of antenna element considered in this project is a microstrip antenna (also known as a patch antenna), since it is intended to be conformally mounted on a smooth surface or a similar device.

Given an array of identical elements, the total array pattern, neglecting coupling, is represented by the product of the single element pattern of the electric field and the array factor [59]. A planar array configuration was chosen because of its ability to scan in three-dimensional (3D) space. For $M \times N$ identical elements with uniform spacing placed on the xy-plane, as shown in Fig. 7.1, the array factor is given by [59]

$$[AF(\theta, \phi)]_{M \times N} = \sum_{m=1}^{M} \sum_{n=1}^{N} w_{mn} e^{j[(m-1)\psi_x + (n-1)\psi_y]}$$

$$\psi_x = \beta d_x (\sin\theta \cos\phi - \sin\theta_0 \cos\phi_0)$$

$$\psi_y = \beta d_y (\sin\theta \sin\phi - \sin\theta_0 \sin\phi_0)$$

(7.1)

where β is the phase constant, w_{mn} represents the complex excitations of the individual elements, and (θ_0, ϕ_0) represents the pair of elevation and azimuth angles, respectively, of maximum radiation. It is the w_{mn}'s and $\psi_{x,y}$'s that the adaptive beamforming algorithms needs to adjust to place the maximum of the main beam toward the (SOI) and nulls toward the SNOIs.

For narrow-beamwidth designs, the main beam can resolve the SOIs more accurately and allow the smart antenna system to reject more SNOIs. Although this may seem attractive for a smart antenna system, it has the disadvantage, because of the large number of elements that may be needed, of increasing the cost and the complexity of the hardware implementation. Moreover, larger arrays require more training bits and hence the overall *throughput* is also affected. Therefore, this tradeoff is examined based on the needs of the network *throughput*, and it has been found that a planar array of 8×8 antenna elements gives the necessary *throughput* for the MANET of this project.

The microstrip array of this project was designed to operate at a frequency of 20 GHz using a substrate material of silicon with a dielectric constant of 11.7 and a loss tangent of 0.04, a thickness of 0.3 mm and an input impedance of 50 Ohms. Using *Ensemble®*, the physical dimensions of the final design of the rectangular patch are listed in Fig. 7.1 and the magnitude of the return loss (S_{11}) versus frequency (*return loss*) is shown as a verification of the design in Fig. 7.2. The E-*plane* and the H-*plane* far-field patterns of a single microstrip element, for the design of Fig. 7.1, are shown in Fig. 7.3.

Using the dimensions of the single patch antenna, a planar array of 8×8 microstrip patches, also shown in Fig. 7.1, with $\lambda/2$ (half-wavelength) interelement spacing (maximum allowable spacing for a well-correlated antenna array) where $\lambda = 1.5$ cm was designed.

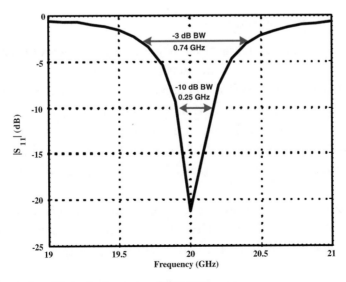

FIGURE 7.2: Return loss (S_{11}) of microstrip of Fig. 7.1.

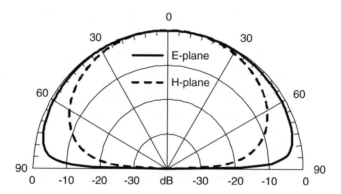

FIGURE 7.3: Single element microstrip patch radiation patterns; E-plane ($\phi = 0°$) and H-plane ($\phi = 90°$).

Once the antenna array design is finalized, the DOA algorithm computes the angle of arrival of all signals based on the time delays. For an $M \times N$ planar array, as shown in Fig. 7.1, these are computed by

$$\tau_{mn} = \frac{md_x \sin\theta \cos\phi + nd_y \sin\theta \sin\phi}{v_o}$$
$$m = 0, 1, \ldots, M-1 \tag{7.2}$$
$$n = 0, 1, \ldots, N-1$$

where v_o is the speed of light in free space.

7.3   MUTUAL COUPLING

The impedance and radiation pattern of an antenna element changes when the element is radiating in the vicinity of other elements causing the maximum and nulls of the radiation pattern to shift. Such changes lead to less accurate estimates of the angles of arrival and deterioration in the overall pattern. These detrimental effects intensify as the interelement spacing is reduced [59, 108–113]. Consequently, if these effects are not taken into account by the adaptive algorithms (beamformer or DOA), the overall system performance will degrade. However, using a mutual coupling matrix (MCM), mutual coupling effects can be compensated [108–113].

To compensate for mutual coupling, a mutual coupling matrix $\mathbf{C}$ is used to revise the updated weight coefficients of the array either in the radiation or receiving mode [113]. The expression for the mutual coupling matrix is given either by [108]

$$\mathbf{C} = Z_L(\mathbf{Z} + Z_L\mathbf{I})^{-1} \tag{7.3}$$

or by [110]

$$\mathbf{C}' = (Z_A + Z_L)(\mathbf{Z} + Z_L\mathbf{I})^{-1} \qquad (7.4)$$

The two are related by $\mathbf{C}' = [(Z_A + Z_L)/Z_L]\mathbf{C}$. In the above two equations, $\mathbf{I}$ is the identity matrix, $\mathbf{Z}$ is the impedance matrix, and Z_L is the load impedance (i.e., 50 Ω). These expressions describe how the individual antenna elements are coupled with one another, which is the information needed to compensate for the mutual effects by the adaptive beamforming algorithm.

7.4   ADAPTIVE SIGNAL PROCESSING ALGORITHMS

The unitary ESPRIT algorithm [176] was chosen as the DOA algorithm for this study. Following the DOA, the adaptive beamformer is introduced to generate the complex excitation weights. The performance of the beamformer over AWGN channels and of the optimal combiner for Rayleigh-fading channels is analyzed.

7.4.1   DOA

After the antenna array receives all the signals from all directions, the DOA algorithm determines the directions of all impinging signals based on the time delays implicitly supplied by the antenna array using (7.2). Then, the DOA algorithm supplies this information to the beamformer to orient the maximum of the radiation pattern toward the SOI and to reject the interferers by placing nulls toward their directions.

The most popular type of DOA algorithms for uniform planar arrays is the ESPRIT. Some of the recent contributions in this area include [124, 176, 177]. In the original version of the ESPRIT algorithm [122], mentioned earlier, only a single invariance is exploited, which is sufficient for estimating DOAs in a single dimension (linear array) but not, in azimuth and elevation angles simultaneously, as needed for planar arrays. Shortly after the development of the first version of ESPRIT, a multiple invariance relation was developed in [178]. This MI-ESPRIT exploits multiple invariances along a single spatial dimension and it is based on the subspace fitting formulation of the DOA problem [179]. The disadvantage of MI-ESPRIT is that it involves the minimization of a complex, nonlinear cost function using an iterative Newton method. The MI-ESPRIT method was extended from the one-dimensional (1D) DOA case to computation of both azimuth and elevation directions in [124] where approximations were used to get a suboptimal solution of the subspace fitting problem. The unitary ESPRIT, presented later in [176] for DOA estimation with uniform rectangular arrays, eliminates the nonlinear optimization and provides a closed-form solution for the azimuth and elevation angles. The algorithm in [124] and the two-dimensional (2D) unitary ESPRIT algorithm focus on computing the azimuth and elevation angles while neglecting to provide a

TABLE 7.1: Signals Used to Test the Smart Antenna System [69]

	SOI		SNOI	
DOA	θ_0	ϕ_0	θ_1	ϕ_1
Case 1	0°	0°	45°	0°
Case 2	30°	45°	60°	45°

good algorithm for computing a basis for the signal subspace. They simply suggest the use of an unstructured eigendecomposition of the data matrix. In [180], Strobach first recognized that the structure of the signal subspace could be exploited to provide more accurate estimates of the signal subspace, which in turn resulted in more accurate DOA estimates. The algorithm that uses this equirotational stack structure of the signal subspace to estimate the DOAs is known as the ES-ESPRIT [181].

In the unitary ESPRIT algorithm for the planar array, the azimuth and elevation angles are computed by stacking the received data vectors and computing a basis for the signal subspace. Next, the least-squares solution of the following two equations of the form

$$\mathbf{K}_{u_1}\mathbf{E}_s\mathbf{\Psi}_u = \mathbf{K}_{u_2}\mathbf{E}_s \quad \text{and} \quad \mathbf{K}_{v_1}\mathbf{E}_s\mathbf{\Psi}_v = \mathbf{K}_{v_2}\mathbf{E}_s \tag{7.5}$$

is obtained. The columns of $\mathbf{E}_s$ contain a basis for the signal subspace and the $\mathbf{K}$ matrices are sparse matrices that depend on the symmetric geometry and size of the array. The $d \times d$ matrices $\mathbf{\Psi}_u$ and $\mathbf{\Psi}_v$ are the rotational operators of the rotational invariance relation and are the solutions to (7.5). The azimuth angles $\mathbf{\Phi}_s$ are obtained from the eigenvalues of $\mathbf{\Psi}_u$ and the elevation angles $\mathbf{\Theta}_s$ from the eigenvalues of $\mathbf{\Psi}_v$. Details of this algorithm can be found in [176].

The unitary ESPRIT algorithm has been implemented as the DOA algorithm for this project. Using the signals of Table 7.1 as input signals to the ESPRIT, it has been observed to give accurate results in the presence of noise and mutual coupling as shown in Table 7.2 [70].

7.4.2 Adaptive Beamforming

Using the information supplied by the DOA, the adaptive algorithm computes the appropriate complex weights to direct the maximum radiation of the antenna pattern toward the SOI and places nulls toward the SNOIs. There are several general adaptive algorithms used for smart antennas [144, 182] and they are typically characterized in terms of their convergence properties and computational complexity. The simplest algorithm is the DMI algorithm where

TABLE 7.2: Esprit Simulation Results [69]

DESCRIPTION		SOI		SNOI	
		θ_0	ϕ_0	θ_1	ϕ_1
Case 1	Without noise	0.000°		45.000°	0.000°
Case 2	Without noise	30.000°	45.000°	60.000°	45.000°
Case 1	AWGN: $\mu = 0$, $\sigma^2 = 0.1$	0.030°		44.945°	0.000°
Case 2	AWGN: $\mu = 0$, $\sigma^2 = 0.1$	30.004°	44.955°	60.060°	44.973°
Case 1	Mutual coupling	0.0508°		44.509°	0.0133°
Case 2	Mutual coupling	30.138°	45.719°	61.072°	45.460°

the weights are computed from the estimate of the covariance matrix [157]. The accuracy of the estimate of this matrix increases as the number of data samples received, allowing more accurate weights to be computed.

The adaptive beamforming algorithm chosen in this project is the LMS for its low complexity [157]. Based on the array geometry of Fig. 7.1, the signals received by the array are given in a matrix form by

$$\mathbf{x} = \mathbf{x}_d + \sum_{i=1}^{L} \mathbf{x}_i + \mathbf{x}_n \qquad (7.6)$$

where $\mathbf{x}_d$ is the desired signal matrix, $\mathbf{x}_i$ is the ith interfering signal matrix and $\mathbf{x}_n$ is the additive noise matrix with independent and identically distributed (*i.i.d.*) complex Gaussian entries with zero mean and variance 0.5 per complex dimension are assumed and L is the number of interferers. Let s_d and s_i denote the desired and the interfering signals, respectively, such that their power is normalized to unity, i.e., $\mathcal{E}\{s_d\}^2 = 1$ and $\mathcal{E}\{s_i\}^2 = 1$. Hence, the received signal vector can be written as

$$\mathbf{x} = \sqrt{\frac{\rho_d}{64}} s_d \mathbf{u}_d + \sum_{i=1}^{L} \sqrt{\frac{\rho_i}{64}} s_i \mathbf{u}_i + \mathbf{x}_n \qquad (7.7)$$

where $\mathbf{u}_d$ and $\mathbf{u}_i$ are the desired and ith interfering signal propagation matrices and ρ_d and ρ_i are the received desired signal-to-noise ratio and ith interference to noise ratio. Note that the received powers are normalized so that they represent the desired SNR.

Arranging the input signals in a column vector $\mathbf{x}_k$, the LMS algorithm computes the complex weights $\mathbf{w}_k$ iteratively using [157]

$$\mathbf{w}_{k+1} = \mathbf{w}_k + \mu\mathbf{x}_k\left(d_k - \mathbf{x}_k^T\mathbf{w}_k\right) \qquad (7.8)$$

where d_k is a sample of the desired signal (i.e., the SOI) at the kth iteration and μ denotes the step size of the adaptive algorithm. In (7.8), μ denotes the step size, which is related to the rate of convergence; in other words, how fast the LMS algorithm reaches steady state. The smaller the step size, the longer it takes the LMS algorithm to converge; this would mean that a longer training sequence would be needed, thus reducing the bandwidth. Therefore, μ plays a very important role in the network *throughput*, as will be discussed later.

7.4.3   Beamforming and Diversity Combining for Rayleigh-Fading Channel

At this point, the performance of adaptive antenna arrays over fading channels is explored. Here, the optimum combining scheme, resulted from the MMSE criterion, is considered in which the signals received by multiple antennas are weighted and summed such that the desired SINR at the output is maximized. The implementation of the optimum combining scheme of [183, 184] has been used to combine the signals. The scheme has been implemented using the LMS algorithm [185]. During the transmission of the actual data, the weights are updated using the imperfect bit decisions as the reference signal, i.e., the LMS algorithm is used in the tracking mode.

In order to simulate the fading channel, a filtered Gaussian model [68] was used with a first-order low-pass filter. The length of the training sequence was again set to 60 symbols but transmitted periodically every 940 actual data symbols (i.e., 6% overhead). The performance of the LMS algorithm over a Rayleigh flat fading channel is presented in Fig. 7.4.

The BER results show that when the Doppler spread of the channel was 0.1 Hz, the performance of the system degraded about 4 dB if one equal power interferer was present compared to the case of no interferers. If the channel faded more rapidly, it was observed that the LMS algorithm performs poorly. For example, the performance of the system over the channel with 0.2 Hz Doppler spread degraded about 4 dB at a BER of 10^{-4} compared to the case when the Doppler spread was 0.1 Hz. An error floor for the BER was observed for SNRs larger than 18 dB. For a relatively faster fading in the presence of an equal power interferer, the performance of the system degrades dramatically implying that the performance of the adaptive algorithm depends highly on the fading rate. Furthermore, if the convergence rate of the LMS algorithm is not sufficiently high to track the variations over rapidly fading channel, adaptive algorithms with faster convergence should be employed.

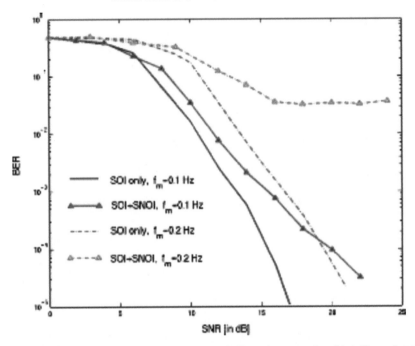

FIGURE 7.4: BER over Rayleigh-fading channel with Doppler spreads of 0.1 Hz and 0.2 Hz for the signals of Table 7.1. The length of the training symbol is 60 symbols and is transmitted every data sequence of length 940 symbols [24].

7.5 TRELLIS-CODED MODULATION (TCM) FOR ADAPTIVE ARRAYS

To further improve the performance of the system, TCM [186] schemes are used together with the adaptive arrays [187–189]. In this scheme, the source bits are mapped to channel symbols using a TCM scheme and the symbols are interleaved using a pseudo-random interleaver in order to uncorrelate the consecutive symbols to prevent bursty errors. The actual transmitted signal is formed by inserting a training symbol sequence to the data sequence periodically. The signal received by the adaptive antenna array consists of a faded version of the desired signal and a number of interfering signals plus AWGN. The receiver combines the signals from each antenna element using the LMS algorithm. During the transmission of the data sequence, a decision directed feedback is used, as it was done in the previous section. The combined receiver output at time k is given by: $r_k = \mathbf{w}_k^H \mathbf{x}_k$ where $\mathbf{w}_k$ and $\mathbf{x}_k$ are the weight vector and received signal vector at time k, respectively. After deinterleaving, the sequence of the combiner outputs $\{r_k\}$ is used to compute the Euclidean metric $m(r_k, \hat{s}_k) = \mathbf{Re}(r_k, \hat{s}_k^*)$ for all possible transmitted symbols $\hat{s}_k$. The set of branch metrics $m(r_k, \hat{s}_k) : \hat{s}_k \in X_q$ is then fed into the Viterbi decoder.

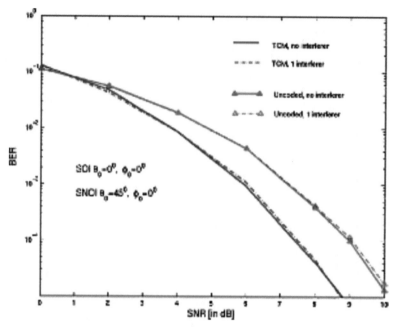

FIGURE 7.5: BER for uncoded BPSK and trellis-coded QPSK modulation based on eight-state trellis encoder over AWGN channel for *Case* 1 of Table 7.1 [24].

A trellis coded QPSK modulation scheme based on an eight-state trellis encoder was considered [70]. In Fig. 7.5, the performance of TCM QPSK systems over a Rayleigh-fading and uncoded BPSK over an AWGN channel are compared for both cases of Table 7.1. The desired and the interfering signals are assumed to be perfectly synchronized, which can be considered as a worst case assumption. It is also assumed that the interfering signals and desired signal have equal power. For the simulation process, the length of the training sequence is also 60 symbols followed by a sequence of 940 symbols at each data frame. It is observed that the adaptive antenna array using the LMS algorithm can suppress one interferer without any performance loss over both an AWGN channel and a Rayleigh-fading channel. However, the impressive feature is that the performance of the TCM system over a Rayleigh-fading channel is even better than that of the uncoded BPSK system over an AWGN channel by about 1.5 dB at a BER of 10^{-5}.

The same system was then analyzed over a Rayleigh-fading channel, and the BER results for Doppler spreads of 0.1 and 0.2 Hz are shown in Fig. 7.6 for both cases of Table 7.1. A training sequence of length 60 symbols, which was periodically sent every 940 symbols of the actual data, with a symbol rate of 100 Hz and interleaver size of 2000 symbols were used. This scheme is comparable with the uncoded BPSK modulation that has the same spectral efficiency. The BER results for the uncoded BPSK scheme over the same

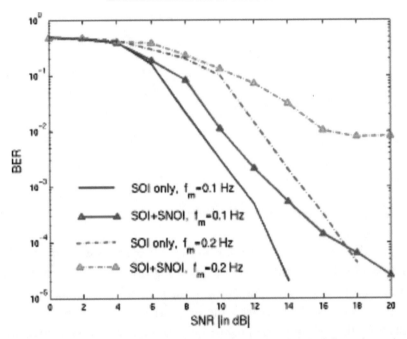

FIGURE 7.6: BER for trellis-coded QPSK modulation over Rayleigh-fading channel with Doppler spreads of 0.1 and 0.2 Hz for both cases of Table 7.1. The length of the training symbol is 60 symbols and is transmitted every data sequence of length 940 symbols [24].

channel are shown in Fig. 7.4. It was observed that when the Doppler spread is 0.2 Hz and there is one interferer, there is still an irreducible error floor on the BER; however, the error floor is reduced compared to the uncoded BPSK case. It can be concluded that the TCM scheme provides some coding advantage in addition to diversity advantage provided by spatial diversity.

7.6   SMART ANTENNA SYSTEMS FOR MOBILE AD HOC NETWORKS (MANETS)

In MANETs there does not exist a fixed network infrastructure and nodes move randomly as shown in Fig. 7.7. Future wireless networks may not be planned and may evolve in an *ad hoc* fashion. In MANETs, data packets are transferred in single hops and the use of directional beams for communication results in reduced interference and hence improved capacity. To facilitate the use of smart antennas in a MANET, nodes must be capable of estimating the direction of the desired node. A few approaches to this problem are suggested in [190] and [191] that use a GPS or the direction of maximum received power. However, with smart

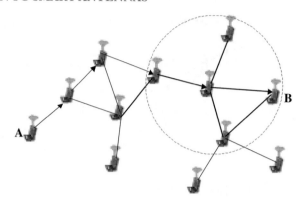

FIGURE 7.7: A typical MANET topology [24].

antennas it is possible to detect the incoming signals using DOA estimation techniques such as MUSIC and ESPRIT algorithms [122, 123] or using LMS-type beamforming algorithms.

The MAC protocol proposed in this work allows nodes to exchange training packets before the data transfer. Nodes start with the isotropic mode of antennas and switch to the directional mode by the end of the training period. Data transfer takes place in the directional mode of antennas. To accomplish this, antennas should be able to operate in both isotropic and directional modes.

7.6.1   The Protocol

The proposed channel access protocol exploits the fact that the interference from a node using directional antennas is low and allows its neighbors to access the channel if the sensed signal power is below a certain threshold. The protocol is based on IEEE 802.11 MAC [192] for TDMA environment, whose details can be found in [193], and is exhibited in Fig. 7.8. It should be emphasized that the introduction of training packets incurs an overhead in the data traffic. If the beamforming algorithms are slow to converge, the required training packet length will be longer, leading to a lower network capacity. Similarly, the antenna parameters, such as the array size and the excitation distribution, influence the capacity. The following section presents some simulation results that show how the capacity of MANETs depends on these parameters.

7.6.2   Simulations

The main objective of the simulations is to qualitatively analyze the capacity improvement in MANETs when smart antennas are used for communication. The simulations also examine the dependence of capacity on various antenna patterns and the length of the training packets. Following are the definitions of the estimated parameters in the simulations:

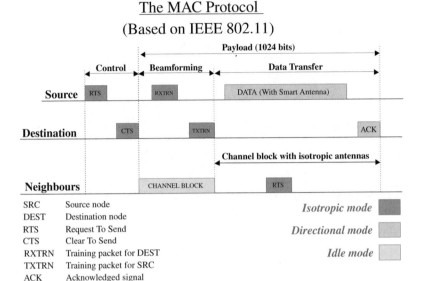

FIGURE 7.8: The proposed channel access protocol [24].

- Average network *throughput* (G_{avg}) is defined as the average number of successfully transmitted packets in the network during a packet time.

- Average load (L_{avg}) is defined as the average number of packets generated in the network during a packet time.

- Average packet delay (T_{avg}) is the average delay experienced by a packet before it is received by the destination.

An *ad hoc* network of 55 uniformly distributed nodes was chosen, as shown in Fig. 7.9.

OPNET Modeler/Radio tool (a simulation software package by *OPNET* Technologies, Inc., used to study, design, and develop communication networks, devices, and protocols) is used to simulate the network. The load at each node is assumed to be Poisson distributed and the mobility is modeled by changing position at random every two packets. The table in Fig. 7.10 shows the values used in simulations for various packet lengths and time intervals specified in the protocol. All packet lengths are normalized to the payload or DATA packet length. Packet lengths of TXTRN and RXTRN are made variable to analyze the performance of the protocol for different training periods.

Network capacity for various antenna patterns is evaluated in order to guide the antenna design for high network capacity. The training packet length is chosen to be 10% of the payload (DATA) length. Average network *throughput* (G_{avg}) is measured for planar arrays of

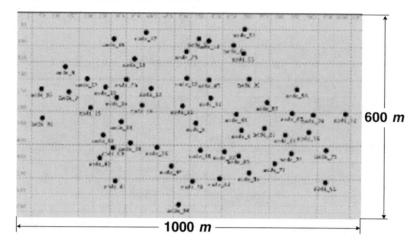

FIGURE 7.9: Network model used for the simulation [24].

Simulation Parameters for MAC

Packet lengths used:

DIFS	0.023 L	
SIFS	0.004 L	
RTS	0.011 L	Control Packets
CTS	0.011 L	
ACK	0.011 L	
TXTRN	Variable	Beamforming Packets
RXTRN	Variable	
DATA	L	Payload (Data)

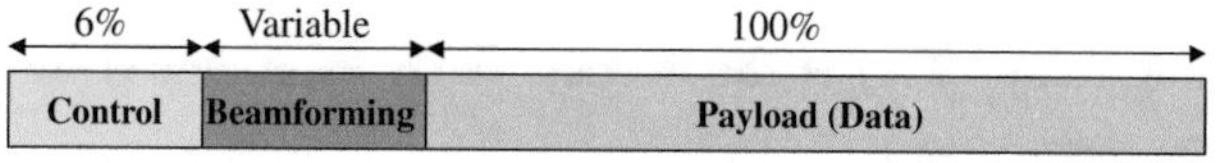

6%   Variable   100%

| Control | Beamforming | Payload (Data) |

FIGURE 7.10: Packet lengths and time intervals used in the protocol simulations [24].

size 8 × 8 and 4 × 4 with Tschebyscheff and Uniform excitation distributions. Fig. 7.11 shows G_{avg} versus L_{avg} for various antenna patterns. The Tschebyscheff arrays were designed for a −26 dB sidelobe level [59]. Neither the uniform nor the Tschebyscheff pattern has been adopted to place a null toward the SNOI. It can be seen that the *throughput* for the case of the 8 × 8 array size is greater compared to the 4 × 4 array size and also the Tschebyscheff arrays provide slightly greater *throughput* than their respective uniform arrays. These can be attributed, respectively, to smaller beamwidths of the 8 × 8 arrays (compared to the 4 × 4 arrays) and lower sidelobes of the Tschebyscheff arrays (compared to the uniform arrays) [59]. In both cases, the smaller beamwidths and lower sidelobes lead to lower cochannel interference.

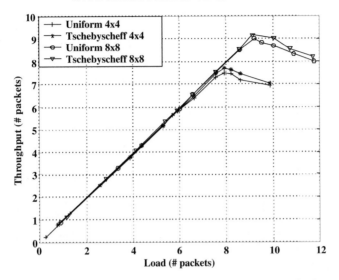

FIGURE 7.11: *Throughput* versus load curves for various antenna patterns [24].

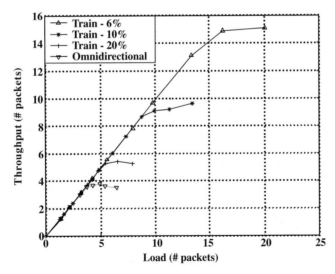

FIGURE 7.12: *Throughput* versus load curves for various training periods [24].

Network capacity for various training packet lengths is evaluated in order to guide the design of beamforming algorithms for high network capacity. Each node is assumed to be equipped with an 8×8 planar array of microstrip patch antennas with Tschebyscheff (-26 dB sidelobes) excitation distribution. Figs. 7.12 and 7.13 show G_{avg} versus L_{avg} and T_{avg} versus L_{avg}, respectively, for the cases when training packet length is 6%, 10%, and 20% of payload using a Tschebyscheff design (-26 dB sidelobes). As can be seen, the network *throughput* is reduced

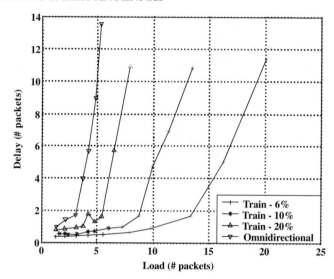

FIGURE 7.13: Delay versus load curves for various training periods [24].

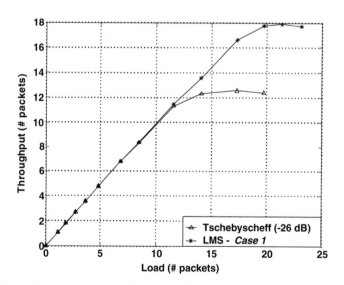

FIGURE 7.14: *Throughput* comparison of a fixed Tschebyscheff pattern with −26 sidelobe level and the pattern for *Case 1* of Table 7.1 [24].

and the packet delays increase rapidly with increasing training packet size. Also, from these figures, it can be observed that the *throughput* of the network is higher when smart antennas are used instead of isotropic antennas.

The network *throughput* is further analyzed using an LMS algorithm generated pattern. This *throughput* is compared in Fig. 7.14 to the *throughput* of a standard Tschebyscheff antenna pattern (-26 dB sidelobes), which does not have an adaptive null toward the SNOI. From this figure, it can be concluded that the adaptive LMS beamforming algorithm leads to higher *throughput* by suppressing the interference (placing a null toward the SNOI) while the Tschebyscheff pattern does not have a null toward the SNOI.

7.7   DISCUSSION

From the results obtained, it is possible to provide certain guidelines for the design of smart antenna systems for optimum capacity in MANETs. Antenna parameters, such as array size and excitation distribution, can be chosen to meet the capacity requirements for a network, based on the simulation results. From these simulation results, it can be concluded that:

(1) radiation patterns with smaller beamwidths result in higher network capacity,

(2) radiation patterns with lower sidelobes can further improve network capacity, and

(3) adaptive radiation patterns (i.e., capable of placing nulls toward the SNOIs) usually produce higher network capacity compared to patterns with lower sidelobes but no nulls toward the SNOIs.

Also, since there is a tradeoff between the network capacity and the training packet length, these simulations assist in choosing a suitable value for the training packet length without compromising on the network capacity. The training period places an upper bound on the convergence speed of the beamforming and DOA estimation algorithms, serving as a guideline for the algorithm design. The results show that training periods greater than 20% reduce the throughput considerably; therefore, it can be inferred that fast beamforming algorithms are critical for high-network capacity.

Employment of smart antenna systems in MANETs creates a wide scope for enhancing the network capacity. Through the design of efficient channel access protocols, spatial diversity of smart antennas can be exploited to increase the capacity of an *ad hoc* network. However, the design of such protocols requires a careful consideration of the system aspects of the smart antenna technology. In this work, a channel access protocol is suggested for MANETs employing smart antennas to communicate. This protocol is built based on the MAC protocol of IEEE 802.11 WLANs [175, 192] for TDMA environment. The protocol facilitates the use of smart antennas and decreases cochannel interference, thereby increasing the capacity of the network.

Finally, it has been shown that in slow fading channels, the performance of the DMI and LMS algorithms is similar. However, in fast fading channels the LMS algorithm is not as effective. Therefore, in such cases, it is suggested to initially use the DMI algorithm in acquisition and then use the LMS algorithm in the tracking mode. Furthermore, the system performance is improved when TCM is combined with antenna diversity.

CHAPTER 8

Space–Time Processing

Space–time processing (STP) has become one of the most investigated technologies in wireless communications as it provides solutions to wireless environment problems such as interference, bandwidth, and range [25]. In this chapter we present the general principles of STP and demonstrate the major benefits from its applications.

8.1 INTRODUCTION

STP signifies the signal processing performed on a system consisting of several antenna elements, whose signals are processed adaptively in order to exploit the rich structure of the radio channel in both the spatial (space) and temporal (time) dimensions. STP techniques can be applied either to the transmitter or the receiver, or both. Fig. 8.1 illustrates different link structures depending on the number of antennas used in receiving or transmitting modes. These options can be associated with both uplink and downlink. Depending on the number of antennas, the channel is classified as *single input* (SI) or *multiple input* (MI) for transmit and *single output* (SO) or *multiple output* (MO).

When STP is applied at only one end of the link, it is usually referred to as a smart antenna technique. When STP is applied at both the transmitter and the receiver, MIMO (*multiple input, multiple output*) techniques are used. Smart antenna and MIMO technologies have emerged as the most promising area of research and development in wireless communications, and they are capable of resolving the capacity limitations due to traffic congestions in future high-speed broadband wireless access networks [25].

It has been recently shown that, under Rayleigh fading, the capacity of a multiple-antenna link increases almost linearly with the number of transmitting antennas provided that there are at least as many receiving antennas as transmitting antennas and the channel gain between each transmitting/receiving antenna pair is known to the receiver [195, 196]. To achieve this intended increase in capacity, various space–time coding schemes have been developed [197, 198]. Fig. 8.2 is an intuitive illustration of this advancement MIMO systems provide. In Fig. 8.2(a) an uplink system is illustrated with multiple antennas at the BS and a single antenna at the MS. The MS radiates omnidirectionally, while the BS is able to adapt its antenna pattern

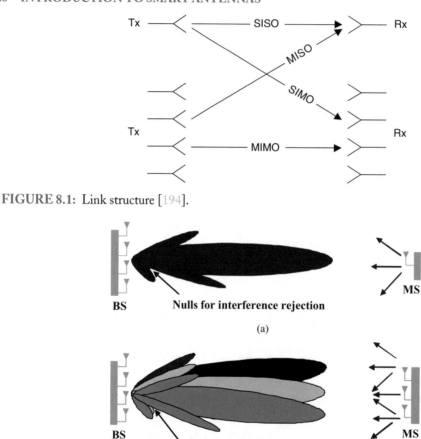

FIGURE 8.1: Link structure [194].

FIGURE 8.2: Uplink antenna systems (a) BS with multiple antennas and MS with single antenna and (b) MIMO system with multiple antennas at both the BS and MS [1].

and focus it into the MS while also rejecting interference through pattern nulls. In a practical scenario, the desired and interfering signals are likely to arrive from many different directions, and therefore the actual beam pattern may appear completely different and not reflect a focusing *spatial filtering* process [1]. In Fig. 8.2(b), a MIMO system is depicted where both the BS and MS are equipped with multiple antennas and several data streams are sent simultaneously over the wireless channel. Each antenna at the MS transmits a different data stream and radiates them omnidirectionally. At the BS, the antenna is capable of forming several beams that can select each of the data streams and correctly receive them. It is clear from this example that the capacity of the system has been significantly increased compared to a conventional system, and it justifies the excitement MIMO systems are generating [28].

Due to the computationally intensive STP algorithms and the limited battery and processing capabilities of handheld mobile devices, until now almost all STP technology development has been related only to base stations and access points. However, with current advancements in low-power mobile device technology and ground-breaking innovation in STP techniques, this technology can also be applied to mobile devices.

Smart antenna technology is an attractive technique that increases spectrum efficiency, range and reliability of wireless networks. Systems that incorporate smart antennas usually have an array of multiple antennas only at one end of the communications link, for example, at the transmit side, such as MISO (*multiple input, single output*) systems; or at the receive side, such as SIMO (*single input, multiple output*) systems. Most conventional smart antenna systems employ the beamforming concept where the signal energy is focused in a particular direction (usually toward the receiver) to increase the received signal-to-noise ratio (SNR). Narrow antenna beams also reduce interference, improving signal to interference noise ratio (SINR) and thereby increasing the efficiency in spectrum management. Other smart antenna schemes improve the link quality by taking advantage of the diversity gain offered by multiple transmitting antennas.

When multiple antenna elements are used, the probability of losing a transmitted signal decreases exponentially with the number of decorrelated signals (or antennas). The diversity scheme used in current SIMO (or MISO) wireless LAN (WLAN) systems incorporates a simple switching network to select, out of an array of two antennas, the antenna that yields the highest SNR. MIMO systems can turn multipath propagation, usually harmful in wireless transmission, into an advantage for increasing the user's data rate.

Diversity-based and smart antenna schemes do not increase the maximum data rate or significantly extend the range of operation; they simply improve the link quality and the efficient use of the spectrum. In contrast, the capacity of MIMO systems, in which antenna arrays are deployed at both the transmitter and the receiver, far exceeds that of conventional smart antennas [25].

In a multipath fading environment, the transmitted signal is reflected by various objects such as walls, buildings, trees and mountains before reaching the receiver. MIMO antenna techniques, accompanied with space–time processing, exploit rich scattering environments by sending independent data streams out of all the transmitting antennas simultaneously and in the same frequency band.

For example, a MIMO-based WLAN 802.11 system with four transmitting and four receiving antennas leads to a fourfold capacity gain up to 216 Mbits/s (4×54 Mbits/s), which can be shared by multiple hotspot users [25]. This type of MIMO technique is referred to as *spatial multiplexing* (SM). Depending on the environmental conditions experienced by the mobile device, the performance improvement of MIMO systems can be applied in two ways.

When the channel conditions and SNR are favorable, the SM technique is used to increase the data rate [25]. In this case, the receiver expends some (if not all, depending on the STP algorithm used) of its degrees of freedom on retrieving the multiple signals rather than providing diversity against fading. However, at longer distances, multiple transmitting and receiving antennas are used to provide diversity and array gain for increased range. Depending on the channel conditions, a link adaptation algorithm, usually residing in the *media-access controller* (MAC) processor, provides the switching between diversity and SM modes of operation. Robust implementation necessitates the ability to adapt to the surrounding environment.

Depending on the propagation channel conditions and STP technique implemented, an N-fold, where N is the number of antennas on the transmitting and receiving ends, MIMO system can yield up to an N-fold capacity increase over that of a single-input, single-output (SISO) system.

When signals are coherently combined at the receiver using techniques such as maximal ratio combining (MRC), the average received SNR increases by $10 \cdot \log_{10}(N)$, where N is the number of receive antennas [25]. Obviously, there is a 6 dB improvement with a four-antenna solution. Fig. 8.3 illustrates the applications and benefits of STP.

In interference-cancelling MIMO systems, it is better to have more receiving antennas than transmitting. For example, if the number of transmitting antennas in a MIMO system is N, in order to cancel one interfering spatial multiplexing user with N independent data

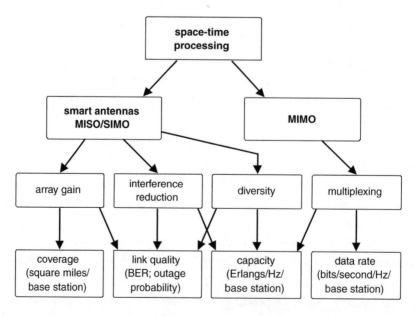

FIGURE 8.3: Space–time processing; applications and benefits [25].

streams, the preferred number of receiving antennas is $2N$. Each interfering multiplexing data stream is seen at the interference-cancelling MIMO receiver as a separate interferer. Therefore, N antennas are used to cancel the interference, and the remaining N antennas are used to demultiplex the desired data streams and achieve diversity gains.

The technique of joint spatial and temporal processing (an overview may be found in [27, 194]) was originally developed for multiuser wireless communications to provide co-channel interference mitigation. Later it was found that space–time processing can also be used to improve SNR, reduce the effect of multipath, provide diversity and increase array gain. In particular, the problem of blind space–time signal processing [199, 200] has gained significant attention in the recent years with the pioneering work on blind equalization using second-order statistics by Tong et al. [201] in 1994 and work on blind signal subspace based methods by Moulines et al. [202] in 1995 (the reader is also referred to [203, 204]). An elegant projection based solution to the multiuser blind equalization problem was proposed by Talwar [205, 206] for the ISI-free channel and Van der Veen [200] for the delay spread channel. The exploitation of the coding dimension in multichannel blind equalization still remains a promising and fertile area of research.

The first evidence of commercially successful small-form-factor multiple antenna technologies can be found in Japan with NTT Docomo's personal digital cellular (PDC) and 3G Foma handsets, as well as in current 802.11 WLAN systems that use two diversity antennas at the receiving end. As described earlier, this technique does not increase the maximum data rate neither significantly extends the range of operation. However, it is a clear proof that multiple antenna technology is steadily penetrating the consumer product market. The biggest challenge to make STP technology commercially feasible is to make it affordable. To do this, both the signal processing algorithms and radio hardware must be implemented in a cost-effective manner. Solutions that can simultaneously integrate these aspects into next-generation silicon will become the key enabling technology for current and future generations of wireless systems. In what follows, an analysis of the space–time signal and channel models is reviewed that provides the necessary tools to examine the basic principles and unique advantages of space–time processing and beamforming afterwards. Finally, in this chapter results from several studies are incorporated which demonstrate the great benefits the wide employment of MIMO systems can yield.

8.2   DISCRETE SPACE–TIME CHANNEL AND SIGNAL MODELS

To proceed with space-time processing, a discrete channel model is considered. This is derived by sampling the received signal in both space and time. The focus is initially drawn on the case of a single user transmitting a modulated signal in a specular multipath environment. At the transmitter, *digital modulation* is performed, a process by which a baseband signal is converted into an RF signal for transmission. Usually, the digital sequence $\{I_k\}$ is linearly modulated by a pulse shaping function $g(t)$ such that the baseband transmitted signal $s(t)$ is represented in the general form

$$s(t) = \sum_{k=-\infty}^{\infty} I_k g(t - kT) \tag{8.1}$$

where T is the symbol period. The data points I_k may come from any *signal constellation* (a set of vectors). For example, with BPSK the possible data symbols are two $\{\pm 1\}$. In other modulations, such as QPSK or QAM, the sequence I_k is complex-valued, since the signal points have a two-dimensional representation. For the reader's interest, the GSM system uses binary signals with GMSK (Gaussian Minimum Shift Keying) modulation for transmission over the air [207] (Ch. 6).

The fundamental function of a channel in signal processing and communications is to relate the transmitted signal to the received version of it [133]. For a baseband transmitted signal $s(t)$, the received signal $x(t)$ can be expressed as the convolution of the *channel impulse response* $h(t, \tau)$ and $s(t)$ as

$$x(t) = \int_{-\infty}^{\infty} h(\tau, t)s(t - \tau)d\tau + n(t). \tag{8.2}$$

The impulse response $h(\tau, t)$ is a function of both the time delay τ introduced by the channel due to multipath propagation and the time t that accounts for the time evolution. Furthermore, additive noise $n(t)$ is incorporated in (8.2). This is by far the most common assumption regarding noise, although other assumptions can be made as in [71, 72, 208].

The previous expression for a single transmit and receive antenna is straightforwardly extended to the case of multiple antennas. For a communication link with N receive and M transmit antennas, the channel can be described by an $N \times M$ matrix $\mathbf{H}(\tau, t)$ of complex baseband impulse responses. The element $H_{ij}(\tau, t)$ of the matrix denotes the impulse response from transmit antenna j to receive antenna i. That is, each receive antenna observes a noisy superposition of the M transmitted signals corrupted by the multipath fading channel. Hence, NM impulse responses are required to characterize this type of of *Multi-Element Antenna* (MEA) or MIMO channel. At each time instance, each row of $\mathbf{H}$ $[H_{i1}, H_{i2}, \ldots, H_{iM}]$ represents the channel's response from the M transmitting to a single receiving element, whereas each column of $\mathbf{H}$ $[H_{1j}, H_{2j}, \ldots, H_{Nj}]$ represents the channel's response from a single transmitting element to the N receiving antenna. The latter is also referred to as the *spatio-temporal signature* induced by the jth transmit antenna across the receive antenna array [209]. In principle, any channel model that accurately includes the spatial dimension can be used to investigate the correlation

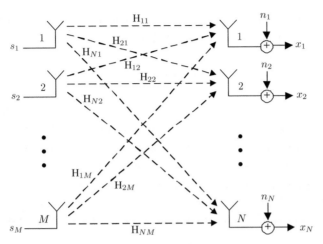

FIGURE 8.4: A wireless link comprising of M transmitting and N receiving antennas [70].

properties of two spatially separated antennas and derive the channel coefficients H_{ij} [133]. An excellent review is found in [210].

In the MEA case, and assuming that the channels between antenna pairs are independent and uncorrelated, the $N \times 1$ vector of received signals $\mathbf{x}(t)$ becomes

$$\mathbf{x}(t) = \int_{-\infty}^{\infty} \mathbf{H}(\tau, t)\mathbf{s}(t - \tau)d\tau + \mathbf{n}(t) \qquad (8.3)$$

where $\mathbf{s}(t)$ denotes the $M \times 1$ vector of transmitted signals and $\mathbf{n}(t)$ is the noise vector of the same length. A representation of the MEA channel is depicted in Fig. 8.4.

Here, it should be stressed that a continuous representation of the signals and impulse responses has been used which naturally arises when deriving the channel model from its physics and electromagnetics prospective [133]. However, most of the recent, and most likely future wireless communication systems employ digital signal processing to a large extent. When devising receiver structures and detector algorithms for these systems, it is more convenient to use a discrete time representation. With the received signal sampled with period T, the notation $\mathbf{x}(n) = \mathbf{x}(nT)$ and

$$\mathbf{x}(n) = \sum_{k=-\infty}^{\infty} \mathbf{H}(k, n)\mathbf{s}(n - k) + \mathbf{n}(n) \qquad (8.4)$$

can be used. Note that $\mathbf{H}$ in (8.4) is the discrete time version of $\mathbf{H}$ in (8.3), as the sampled versions of the transmitted signal and the noise, denoted by $\mathbf{s}(n)$ and $\mathbf{n}(n)$, respectively, are further considered. This is the normal notation used in most of the literature, although the formulation results in some abuse of the notation. For narrowband systems, where the channel is considered to be *frequency-flat*, the main part of the received energy arrives at essentially the same time and the model may be further simplified to

$$\mathbf{x}(n) = \mathbf{H}(n)\mathbf{s}(n) + \mathbf{n}(n). \tag{8.5}$$

Here the channel model reduces to complex matrices comprising complex scalars that relate the received signals of each element to the corresponding transmitted signal from each antenna through a simple multiplying *transfer matrix* which encompasses the entire channel behavior. Further, if the channel is assumed to be time-invariant, the time dependency of the channel may be dropped, i.e. $\mathbf{H}$. For this narrowband MIMO channel matrix, different normalizations have been used in the literature, where the *Euclidean* or *Frobenius* norm $\|\mathbf{H}\|_2 = \sqrt{\sum_{i=1}^{N} \sum_{j=1}^{M} |H_{ij}|^2}$ appears to be the most common one [211].

For the case of Rayleigh flat fading, a simple channel model assumes a circular disc of uniformly distributed scatterers placed around the mobile. In Fig. 8.5 a simple illustration of the scatter disc and the orientation of the mobile and base station are shown. Based on this model, the entries of the channel matrix in (8.5) are generated as follows. Assuming P scatterers S_p, $p = 1, 2, \ldots, P$, are uniformly distributed on a disc of radius R centered around the mobile, the channel coefficient H_{ij} connecting the jth transmit to the ith receive antenna is given by

$$H_{ij} = \sum_{p=1}^{P} a_p \exp\left[-j\frac{2\pi}{\lambda}\left(D_{B_j \to S_p} + D_{S_p \to M_i}\right)\right] \tag{8.6}$$

where $D_{B_j \to S_p}$ and $D_{S_p \to M_i}$ are the distances from the j antenna of the base station to scatterer p, and scatterer p to the i antenna of the mobile unit, respectively. Also, a_p is the scattering

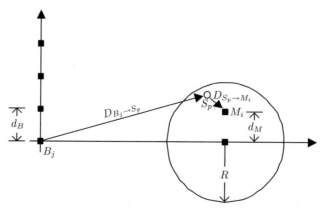

FIGURE 8.5: Geometry of a channel [133].

coefficient from scatterer p and is modeled as a normal complex random variable, with zero mean and unit variance.

For wideband signals (with bandwidth greater than the coherence bandwidth) a different approach must be followed. In essence, an accurate model must account for the replicas of the same signal that arrive at the receiver at different time instances. Equivalently, assuming zero propagation time, each received signal depends on the consecutive signals transmitted in a time window from $L - 1$ previous sampling periods until the current period. We want to express this model as

$$\mathbf{x}(n) = \tilde{\mathbf{H}}(n)\tilde{\mathbf{s}}(n) + \mathbf{n}(n) \tag{8.7}$$

where the matrix $\tilde{\mathbf{H}}$ is $N \times ML$ and the vector $\tilde{\mathbf{s}}$ is $ML \times 1$. The parameter L introduced is responsible to capture almost all the signal energy as it arrives at different delays and is given, as already stated, by the ratio of the *delay spread* (dispersion of the channel in the time domain) to the symbol duration ($L = \lceil \frac{T_m}{T} \rceil$). Matrix $\tilde{\mathbf{H}}$ can be written as

$$\tilde{\mathbf{H}} = \begin{bmatrix} \tilde{\mathbf{h}}_{11} & \tilde{\mathbf{h}}_{12} & \dots & \tilde{\mathbf{h}}_{1M} \\ \tilde{\mathbf{h}}_{21} & \tilde{\mathbf{h}}_{22} & \dots & \tilde{\mathbf{h}}_{2M} \\ \vdots & \vdots & \ddots & \vdots \\ \tilde{\mathbf{h}}_{N1} & \tilde{\mathbf{h}}_{N2} & \dots & \tilde{\mathbf{h}}_{NM} \end{bmatrix} \tag{8.8}$$

where each element $\tilde{\mathbf{h}}_{ij}$ is a row vector of length L containing the channel impulse responses from the transmit antenna j to receive antenna i from the present to $L - 1$ previous instances and expressed as

$$\tilde{\mathbf{h}}_{ij} = \left[h_{ij,n}, h_{ij,n-1}, \dots, h_{ij,n-(L-1)} \right].$$

Similarly, vector $\tilde{\mathbf{s}}$ can be written in block form as

$$\tilde{\mathbf{s}} = [\mathbf{s}_1, \mathbf{s}_2, \dots, \mathbf{s}_M]^T \tag{8.9}$$

where $\mathbf{s}_j$ is a column vector of length L containing the transmitted symbols by the j antenna from the present to $L - 1$ previous instances and written as

$$\mathbf{s}_j = \left[s_{j,n}, s_{j,n-1}, \dots, s_{j,n-(L-1)} \right]^T.$$

Another useful representation of the received signal $\mathbf{x}(n)$ in (8.7) is

$$\mathbf{x}(n) = \sum_{j=1}^{M} \hat{\mathbf{H}}_j \mathbf{s}_j(n) + \mathbf{n}(n) \tag{8.10}$$

where the $N \times L$ matrix $\hat{\mathbf{H}}_j$ denotes the channel response from the jth transmit antenna to the N receive antennas for the L ISI symbols, expressed as

$$\hat{\mathbf{H}}_j = \left[\tilde{\mathbf{h}}_{1j}^T, \tilde{\mathbf{h}}_{2j}^T, \ldots, \tilde{\mathbf{h}}_{Nj}^T \right]^T, \quad j = 1, 2, \ldots, M. \tag{8.11}$$

8.3   SPACE–TIME BEAMFORMING

In previous chapters, we examined only space combining where the $N \times 1$ space-only vector $\mathbf{w}$ is applied to the $N \times 1$ vector of signals at the receive antenna elements (a single complex weight assigned to each antenna) resulting in the usual output

$$y(n) = \mathbf{w}^H \mathbf{x}(n). \tag{8.12}$$

Space-only processing works best if each antenna element is provided a signal with the same time dispersion, *e.g.* the same shape of the impulse response [212]. However, this is not true in general. In a multipath environment, which is usually the case, the received power level is a random function of the user's location and time's evolution depending on the occurred fading. On the other hand, a separate equalization for each antenna was performed to combat multipath propagation, before spatially combining the signals, would be optimal only for the case that at each delay the multipath components arrive from the same direction [212]. In general, this is not true either.

A reasonable solution is to apply a joint *space-time filter* or equalizer in order to take advantage of processing in two dimensions rather than one. The space-time combining is a direct generalization of the space-only combining. The combiner is assumed to have K time taps. Each tap denoted by $\mathbf{w}(i)$, $i = 0, 1, \ldots, K - 1$, is an $N \times 1$ weight vector defined as above. The output of the space-time beamformer is expressed as [26]

$$y(n) = \sum_{i=0}^{K-1} \mathbf{w}^H(i) \mathbf{x}(n - i) \tag{8.13}$$

or in matrix form [26]

$$y(n) = \mathbf{W}^H \mathbf{X}(n) \tag{8.14}$$

where $\mathbf{W}$ and $\mathbf{X}(n)$ are $KN \times 1$ vectors with

$$\mathbf{W} = \left[\mathbf{w}^H(0), \mathbf{w}^H(1), \ldots, \mathbf{w}^H(K - 1) \right]^H \quad \text{and} \tag{8.15a}$$
$$\mathbf{X}(n) = \left[\mathbf{x}^T(n), \mathbf{x}^T(n - 1), \ldots, \mathbf{x}^T(n - K + 1) \right]^T. \tag{8.15b}$$

Fig. 8.6 depicts the structure of the space-time beamformer.

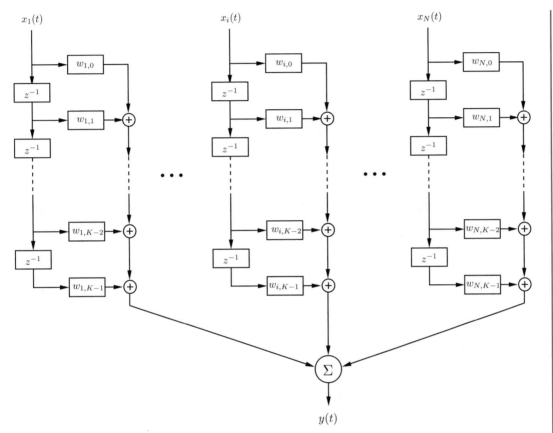

FIGURE 8.6: Structure of the space-time beamformer [20].

8.4   INTERSYMBOL AND CO-CHANNEL SUPPRESSION

The data models developed provide the necessary tools to demonstrate this powerful ability of space-time processing, the simultaneous ISI and CCI suppression. To simplify our present analysis, we introduce a similar in concept system in which Q co-channel users, each one equipped with a single antenna, are present, rather than a single user with M transmit antennas, while the antenna array at the base-station still consists of N elements. According to the assumed model, the received signals at the N elements for an interval of K time taps due to the jth signal transmitting source can be written as a $KN \times 1$ space-time vector [26]

$$\mathbf{X}_j(n) = \mathcal{H}_j \mathbf{S}_j(n) + \mathbf{N}_j(n) \tag{8.16}$$

where $\mathbf{S}_j(n) = \left[s_j(n), s_j(n-1), \ldots, s_j(n-K-L+2) \right]^T$ and $\mathcal{H}_j$ is a $KN \times (K+L-1)$

channel matrix given by [26]

$$
\mathcal{H}_j =
\begin{bmatrix}
\hat{\mathbf{H}}_{j,n} & \mathbf{0} & \cdots & \mathbf{0} \\
\mathbf{0} & \hat{\mathbf{H}}_{j,n-1} & \ddots & \vdots \\
\vdots & \ddots & \ddots & \mathbf{0} \\
\mathbf{0} & \cdots & \mathbf{0} & \hat{\mathbf{H}}_{j,n-K+1}
\end{bmatrix}
\tag{8.17}
$$

where $\mathbf{0}$ is an $N \times 1$ column vector of zeros.

The $\mathcal{H}_j$ matrix has a block Toeplitz structure (has the same elements along diagonals) which stems from the linear time-invariant convolution operation with the symbol sequence [26]. Assuming temporarily a noise-free scenario the output of the linear space-time combiner due to the jth source is given by [26]

$$
y_j(n) = \mathbf{W}_j^H \mathcal{H}_j \mathbf{S}_j(n).
\tag{8.18}
$$

With Q users present transmitting towards the N-element base station, the output of the space-time receiver is generalized to

$$
y(n) = \sum_{j=1}^{Q} \mathbf{W}_j^H \mathcal{H}_j \mathbf{S}_j(n).
\tag{8.19}
$$

8.4.1   ISI Suppression

In general, the purpose of the linear filter $\mathbf{W}_j$ is to perform channel equalization to compensate the effects of ISI for the jth user in the absence of CCI [26]. Tutorial information on channel equalization can be found in [71, 213]. To suppress ISI for the jth user, the convolution product between $\mathbf{W}_j$ and the channel responses must satisfy the following so-called zero-forcing condition

$$
\mathbf{W}_j^H \mathcal{H}_j = [0, 0, \ldots, 0, 1, 0, \ldots, 0, 0].
\tag{8.20}
$$

The location of the entry "1" above represents the delay of the combined channel-equalizer impulse response. From an algebraic point of view, $\mathcal{H}_j$ should have more rows than columns, or $KN \geq (K + L - 1)$, for such solutions to exist [26]. Therefore, it is of great importance that the space-time filter to have large number of time taps K (degrees of freedom) to allow for ISI suppression. An obvious selection is to keep the number of time taps K at least equal to the number of distinct multipaths L.

8.4.2   CCI Suppression

The purpose of CCI suppression in a cellular communications systems is to identify and enhance the contributing of a singledesired jth user while ideally suppressing all others. This can be achieved by enforcing *orthogonality* between the response of the space-time combiner and the response of the channel of the users to be rejected [26]. In mathematical form this is expressed as [26]

$$\mathbf{W}_j^H \boldsymbol{\mathcal{H}}_f = \mathbf{0}^T, \quad \text{for all } f \neq j \in [1, 2, \dots, Q] \tag{8.21}$$

where $\mathbf{0}$ is a $(K + L - 1) \times 1$ column vector of zeros. In (8.21), there are $(Q - 1)(K + L - 1)$ scalar equations to be solved. Once more, the number of unknowns is KN (the size of $\mathbf{W}_j$). Simultaneously, the receiver must capture the energy transmitted by the desired user and thus, $\mathbf{W}_j^H \boldsymbol{\mathcal{H}}_j \neq \mathbf{0}^T$. This condition requires that $\boldsymbol{\mathcal{H}}_j$ and $\{\boldsymbol{\mathcal{H}}_f\}_{f \neq j}$ should not span the same column subspace which is satisfied by the meaningful assumption that the desired signal and interference do not arrive from the same direction [27].

8.4.3   Joint ISI and CCI Suppression

To completely recover the signal transmitted by one desired user in the presence of ISI and CCI, both channel equalization and separation is required. The output of the space-time combiner which satisfies both (8.20) and (8.21) is written as [26]

$$\mathbf{W}_j^H \widetilde{\boldsymbol{\mathcal{H}}} = [0, 0, \dots, 0, 1, 0, \dots, 0, 0] \tag{8.22}$$

where $\widetilde{\boldsymbol{\mathcal{H}}} \triangleq [\boldsymbol{\mathcal{H}}_1, \dots, \boldsymbol{\mathcal{H}}_{j-1}, \boldsymbol{\mathcal{H}}_j, \boldsymbol{\mathcal{H}}_{j+1}, \dots, \boldsymbol{\mathcal{H}}_Q]$ is $KN \times Q(N + L - 1)$ matrix. For existence of the solution for the joint ISI and CCI suppression problem, the multiuser channel matrix $\widetilde{\boldsymbol{\mathcal{H}}}$ needs to have more columns than rows, or $KN \geq Q(N + L - 1)$. It is again critical that smart antennas provide a sufficient number of degrees of freedom [26]. Ideally, if the global channel matrix $\widetilde{\boldsymbol{\mathcal{H}}}$ is full-column rank, then recovery of any particular user is feasible through space-time beamforming.

In practice, however, such ideal joint ISI-CCI reduction is limited by the presence of noise and the lack of perfect synchronization. In the literature, space-time algorithms have been proposed for CCI and ISI rejection for time varying channels. In [214], for example, a space-time algorithm for CCI and ISI reduction in GSM/DCS systems is proposed. Through optimization of a suitable cost function for separable space-time channels, the temporal channel for the Viterbi receiver and the beamformer weights are estimated jointly.

8.5   SPACE–TIME PROCESSING FOR DS-CDMA

Direct-sequence CDMA (DS-CDMA) systems have already a significant penetration into the cellular market with great promising potentials in the future due to their intrinsic advantages over earlier access techniques such as time-division multiple access (TDMA) and frequency-division multiple access (FDMA) [215]. However, it has fundamental difficulties in a scenario whereby the received signal energies are dissimilar (*near-far* environments); the conventional detector is unable to demodulate reliably the weak signals since the transmission power from one user overwhelms signals of the others [216, 217].

Adaptive antenna arrays provide an alternative means to cope with the near-far problem. By steering beams toward the desired user and decreasing the total power level of *multiuser access interference* (MUAI), system near-far resistance, i.e. immunity of the desired user's performance to power variations of the others, can be considerably strengthened [217]. Besides alleviating the near-far problem, antenna arrays also increase the capacity of CDMA systems through interference suppression. This is because the system capacity is limited by interferences, instead of the bandwidth as in TDMA [218, 219], and the reduction in system noise floor due to spatially selective transmission and reception leads to a direct increase in capacity [220].

Recently, there has been an increasing interest in the use of 2D RAKE receivers to simultaneously exploit space and time diversities by combining adaptive antennas with RAKE receivers. In principle, a 2D RAKE receiver allows constructive combination of multipath signals received by an array of antennas while minimizing the MUAI's contribution [217]. This powerful combination provides an optimum output *signal-to-interference-plus-noise* ratio (SINR) for the desired user. The potential of the 2D RAKE has been evaluated and proved by various studies [220–222].

For L resolvable paths and N-element antenna array, the space–time RAKE consists of a beamformer for each path with weights $\mathbf{w}_{ln}$ followed by a standard RAKE. The beamformer is an MMSE type and is followed by a RAKE combiner which does not represent an MMSE time receiver. Fig. 8.7 illustrates a space–time RAKE. Such a receiver leads to an improved figure-of-merit that can be traded for improved coverage or capacity. This improvement results from reducing the intercell CCI through beamforming and ISI by coherently combining resolvable paths.

8.6 CAPACITY AND DATA RATES IN MIMO SYSTEMS

MIMO systems, enabled by the employment of advanced space–time processing techniques, offer a significant increase in spectral efficiency based on the utilization of space diversity both at the transmitter and the receiver. Before finishing this chapter it is considered useful to review and compare the capacity expressions and data rates for SISO/MIMO channels.

With a MIMO system, a high-rate bit stream at the transmitter is decomposed into independent bit sequences, which are then transmitted simultaneously using multiple antennas. Each substream is then encoded into channel symbols. Commonly, the same data rate is imposed

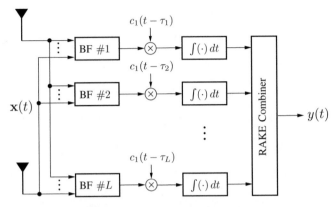

FIGURE 8.7: The space–time RAKE receiver model for CDMA uses a beamformer to spatially separate the signals, followed by a conventional RAKE [20].

on all transmitters, but adaptive modulation rate can also be utilized on each of the substreams [20]. The signals are launched and naturally mixed together in the wireless channel as they use the same frequency spectrum. Receive and transmit antennas must be sufficiently separated in space and/or polarization to create independent propagation paths. At the receiver, after having identified the mixing channel matrix through training symbols, the individual bit streams are recombined to provide the enhanced data rate signal. This transmission scheme results in a linear increase in spectral efficiency compared to a logarithmic increase in more traditional systems utilizing receive-diversity or no diversity. An explanation for this great improvement is the fact that the data stream from each transmitter appears highly uncorrelated to each receiver due to the the rich scattering environment. The general functionality of a MIMO system is shown in Fig. 8.8.

8.6.1   Single-User Data Ratec Limits

The *channel capacity* is a measure of the maximum information that can be transmitted through the channel and received with negligible error probability. With a single transmit antenna and a single receive antenna, the single user data rate bound can be expressed using Shannon's universal equation

$$C = B \log_2 \left(1 + \frac{P_T}{\sigma^2} |h|^2 \right) \quad \text{[bits/s]} \qquad (8.23)$$

where P_T is the total radiated power and σ^2 is the white Gaussian noise power within the channel bandwidth B and $|h|^2$ is the (instantaneous) channel power gain. This expression simply provides an upper bound on channel capacity and only a fraction of which is attainable

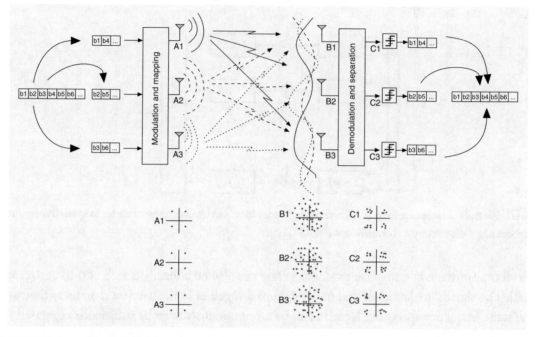

FIGURE 8.8: A basic MIMO scheme with three transmit and three receive antennas yielding threefold improvement in system capacity [223].

since unrealistic conditions are assumed, where the entire system bandwidth is allocated to a single user, no other users are activated and any interference is totally suppressed [28].

In transmit diversity, antenna arrays are only deployed at the transmitter. In this case, base-station antennas must be placed sufficiently appart[1] so that the signals are basically uncorrelated. The upper data-rate bound is given by

$$C = B \log_2 \left(1 + \frac{P_T}{M\sigma^2} \sum_{m=1}^{M} |h_m|^2 \right) \ \text{[bits/s]} \tag{8.24}$$

with h_m being the channel response from each uncorrelated base-station antenna to the single terminal antenna.

8.6.2   Multiple-Users Data Rate Limits

When multiple antennas are used at the transmitter and/or receiver, Shannon's equation, with no instantaneous channel information at the transmitter, can be generalized by [133, 196]

---

[1]The necessary spacing depends on the angle spread, but for typical values is of the order of 5–10 wavelengths.

$$C = B \log_2 \left[\det \left(\mathbf{I}_N + \frac{P_T}{M\sigma^2} \mathbf{H}\mathbf{H}^H \right) \right]$$
$$= \sum_{k=1}^{n} \log_2 \left(1 + \frac{\rho}{M} \frac{\lambda_k}{n} \right) \quad [\text{bits/s}]$$

(8.25)

where $\mathbf{I}_N$ is the identity matrix, $|h|^2$ is replaced by $\mathbf{H}\mathbf{H}^H$ and $\frac{P_T}{\sigma^2}$ by ρ (for convenience), $n = \min(M, N)$, and λ_k are the eigenvalues of $\mathbf{H}\mathbf{H}^H$. The entries of the matrix $\mathbf{H}$ represent h_{nm} independent channel coefficients between the M transmit and N receive antennas. Note that this expression assumes that the available transmit power P_T is uniformly allocated to the M transmit antennas, which is the practical approach when no knowledge of the channel is available at the transmitter. For a wideband MIMO channel the overall channel capacity is given by

$$C = \frac{B}{L} \sum_{l=0}^{L-1} \log_2 \left[\det \left(\mathbf{I}_N + \frac{\rho}{MS} \mathbf{H}_l \mathbf{H}_l^H \right) \right] \quad [\text{bits/s}]$$

(8.26)

with L frequency-flat channels in parallel. The attractiveness and capabilities of MIMO systems are well demonstrated through (8.25) by considering, as an example, the ideal case of $\mathbf{H} = \mathbf{I}_N$ (i.e. equal number of transmitting and receiving antennas with ideal, parallel link between them; no criss-crossing). From (8.25) we get [196]

$$C = BN \log_2 \left(1 + \frac{\rho}{N} \right) \rightarrow \frac{B}{\ln 2} \rho \quad \text{as } N \rightarrow \infty.$$

(8.27)

Unlike in (8.23), capacity scales linearly, rather than logarithmically, with increasing of SNR which demonstrates the significant advantage in using parallel transmission. However, any possible advantage through parallelism, offered with deployment of MEAs, must be carefully assessed since the signal component traversing different paths can strongly interfere [196].

To demonstrate the unprecedented levels of performance unleashed by the simultaneous deployment of base station and terminal arrays, parts of the results in [28] are incorporated with the kind courtesy and permission of the authors. In [28], the figure-of-merit is the idea of *outage rate*, which is the value of C supported with certain (high) probability. The authors [28] choose 90% as the probability of support for their results which implies that 10% of the burst or coding blocks may contain errors. This appears to be reasonable operating point for many applications, although other operating points are certainly possible. For their system, the authors concentrate on the downlink only in a cellular system comprising fairly large cells with every cell partitioned into 120° sectors. They also consider a propagation scenario based on the existence of an area of local scattering around each terminal but with little or no local scattering

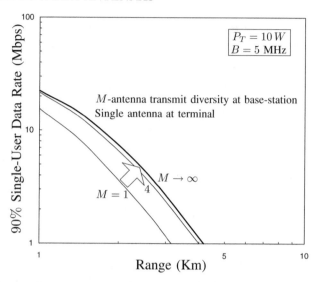

FIGURE 8.9: Single-user data rate supported in 90% of locations vs. range with a directive array at the base-station and a single omnidirectional antenna at the terminal. M is the number of 15 dBi antennas at the base-station; transmit power $P_T = 10W$ and bandwidth $B = 5$ MHz [28].

presumed around the base station. The analysis was conducted in the 2 GHz frequency range and with available bandwidth $B = 5$ MHz. The total transmit power was set to $P_T = 10$ W. Each individual base-station antenna has a gain of 15 dBi whereas the terminal is equipped with a single omnidirectional antenna. Finally, the well-established $COST231$ model, based on Okumura-Hata model [50], is chosen to account for the range-dependent component. Fig. 8.9 shows the single antenna results when M antennas are employed at the base-station. Although there is no fundamental bound in the size of the array, there is little advantage in increasing it beyond $M \approx 3, 4$ because of the diminishing returns associated with adding additionally diversity branches to an already diverse link.

The attention then is turned to systems with both transmit and receive arrays. Based on this model, a class of layered space–time architectures has been proposed and labeled BLAST [224]. In BLAST, multiple data streams are simultaneously radiated using different antennas with a transmit array. With sufficient multipath, a receiver also equipped with an array can separate and successfully decode all the data streams using advanced signal processing techniques that bridge the gap between array processing and multiuser detection. A critical feature of BLAST is that the total radiated power is held constant irrespective of the number of transmit antennas.

As in the transmit diversity case, base-station antennas must be sufficiently spaced apart for proper decorrelation [225]. The receiver is equipped with its own array, set to have equal

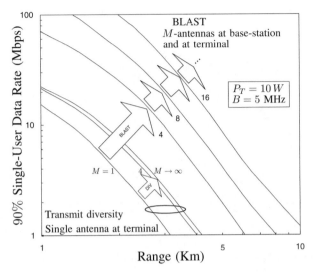

FIGURE 8.10: Single-user data rate supported in 90% of locations vs. range with a directive array at the base-station and a single omnidirectional antenna at the terminal. M is the number of 15 dBi antennas at the base-station as well as the number of omnidirectional antennas at the terminal ($M = N$); transmit power $P_T = 10$ W and bandwidth $B = 5$ MHz [28].

number of elements to that of the base-station array ($M = N$). The capacities, directly derived from (8.25), are depicted in Fig. 8.10.

8.6.3   Data Rate Limits Within a Cellular System

The analysis in [28] is further extended to evaluate the user data rate limits in more realistic conditions examining an entire cellular system. A time-multiplexed system with a 19-cell hexagonal grid is assumed for conduction of Monte-Carlo simulations: a central cell wherein the statistics are collected surrounded by two tiers of interfering cells. Within the same cell, users are uniformly distributed and ensured to be mutual orthogonal such that interference arises exclusively from other cell. For each cell, 120° perfect sectorization is assumed and the user is connected to the sector from which it receives the strongest signal. Finally, rate adaptation and no power control is assumed to resemble actually 3G data systems. The simulation parameters are summarized in Table 8.1.

Fig. 8.11 displays cumulative distributions of system capacity (in megabits per second per sector) over all locations with transmit arrays only and with transmit and receive arrays, as well. The results can be also interpreted as user peak rates (in megabits per second) when the entire capacity is allocated to an individual user. With transmit arrays only, there is a small benefit only in the lower tail of the distribution corresponding to users in harmful locations.

TABLE 8.1: System parameters [28]

MULTIPLEXING	TIME-DIVISION
Sectors per cell	3
Base station antennas	120° perfect sectorization
Terminal antennas	Omnidirectional
Frequency reuse	Universal
Propagation exponent	3.5
Log-normal shadowing	8 dB
Fading	Rayleigh (independent per antenna)
Power control	No
Rate adaptation	Yes
SNR in 90% of locations	≥ 25 dB

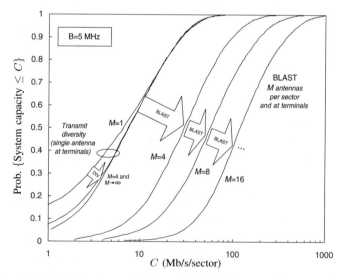

FIGURE 8.11: Cumulative distributions of system capacity with transmit arrays as well as with transmit and receive arrays [28].

The benefits in average and peak system capacity are negligible with increase in the number of transmit arrays. However, the use of additional receive antennas improves dramatically the capacity and data rates of the individual user.

8.6.4   MIMO in Wireless Local Area Networks

The vast spread of Wireless Local Area Networks (WLANs) in our days attracts the use of multiple antennas at both transmitter and receiver ends to establish links with improved quality and even higher data rates. The terminal in such WLANs could be a laptop computer or handheld computers giving opportunity to carry multiple antennas. Examples of WLAN systems are the IEEE 802.11 standard and HiperLAN/2 where standardization is ongoing. Versions of both systems today offer rates of more than 50 Mbits/s with a single terminal antenna. Nevertheless, extremely good Signal-to-Noise-ratios are required to achieve this goal. Multiple antennas on the terminal side are believed to increase the rate further [226] and to relax the SNR requirements. Furthermore the instantaneous link rate may be very high to support large file downloads.

The mentioned WLAN systems both have versions for operation in the 5 GHz band. The propagation characteristics at this frequency are very appropriate for Radio LANs, and it is possible to cover many users at a low cost. It is also possible to use such systems in both indoor and outdoor environments and thus, provide coverage of a hot-spot area such as a campus or an airport. The results shown here are focused on the measured and simulated link capacity at 5.8 GHz (WLAN) and reproduced from [29] with the kind courtesy and permission of its authors. The measurements were carried out in a typical office environment at Telia Research in Malmö, Sweden. The purpose of this investigation was to examine how "rich scattering" in a normal office environment may be, and thus what rate improvement may be possible.

The general planning of the floor consists of office rooms, open spaces and corridors. Most spaces are separated by walls, while glass is used in some of them. The transmitter was positioned in one of the offices, while the receiver was in an open area. The measured channel analyzed was a typical *non-line-of-sight* (NLOS) situation and the distance from transmitter to receiver was 10–15 m. The measurements were made at 5.8 GHz carrier frequency and the transmitter and receiver bandwidth was 400 MHz. By sending a pseudo-noise sequence at the transmitter and correlating with the same synchronous pseudo-noise sequence at the receiver, complex impulse responses were measured. The data was measured with a synthetic antenna array, using one receive and one transmit antenna, both of monopole type. The transmit antenna was moved between seven different positions separated by 300 mm; that is about 6λ (of these seven positions, three were used herein). For each of the seven transmitter positions, the receive antenna was moved between 21 different positions, using a step motor on a track (distance between two adjacent positions was about 13 mm, i.e. about $\lambda/4$). This corresponds to spatial measurements over about 5λ. At each combination of transmit and receive positions 20 samples were taken. All measurements were performed during stationary conditions at night, and the measurement noise was assumed to be very low [29].

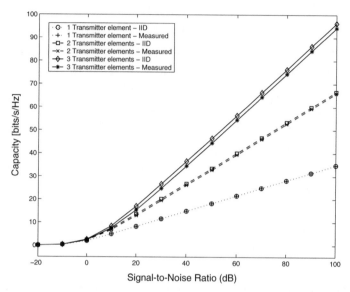

FIGURE 8.12: Capacity for different SNR, with 3 receive elements and one two and three transmit elements, respectively [29].

The capacity for different signal-to-noise ratios, with three receive elements and one, two and three transmit elements is shown in Fig. 8.12. The increase in capacity as the number of transmit elements increases from one to three is indeed substantial. As a reference, the simulated curve for the IID (*independently* and *identically distributed*) Gaussian Channel is plotted as well, indicating low correlation among the elements of the channel matrix **H** under these conditions. The following results are given for the SNR fixed at 20 dB.

Increasing the number of receive and transmit antennas would also increase the capacity. This increase is shown in Fig. 8.13. The capacity increase is large when going from one to the number of elements in the transmit array $M = 3$, while the increase is less when further increasing the number of elements on the receive side. As expected, the increase for $M > N$ follows a logarithmic curve due to SNR increase from noise averaging over the channels.

A final consideration in [29] was that of inter-element spacing at the receive array. The transmit antennas were set to three and experiments were conducted for different number of elements as their inter-element distance is increasing. The resulted capacity is shown in Fig. 8.14. In Fig. 8.15 the capacity dependence of the inter-element distances is shown for two and three receive elements and three transmit elements. Experimental results show that the capacity increase is small when increasing the distance between the elements beyond $\Delta = \lambda$. After comparing the increase to simulations with random IID channels, which provides a

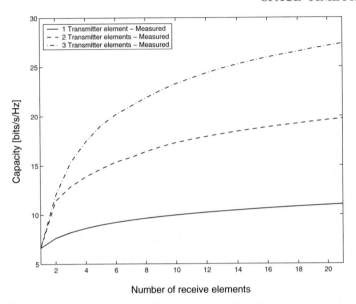

FIGURE 8.13: Capacity as a function of number of receive elements for one, two and three transmit elements, respectively, at an SNR of 20 dB [29].

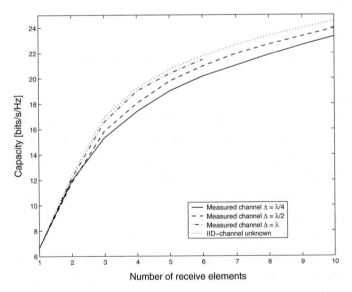

FIGURE 8.14: Capacity dependence on the number of elements in the receive array for different element distances. The SNR is 20 dB and three elements are used in the transmit array [29].

statistical upper bound for the channel capacity, it was concluded that for this experimental setup the subchannels are close to being uncorrelated when the interelement distance was about 2λ.

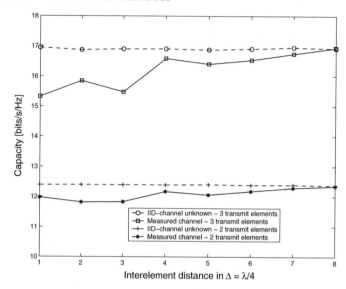

FIGURE 8.15: Capacity dependence on the interelement distance at the receiver, for the measured channel compared to the simulated IID channel. Two and three receive elements are used together with three transmit elements at SNR = 20 dB [29].

8.7   DISCUSSION

In this chapter an attempt was made to illustrate the vast improvements in the capacity and data rates space–time processing techniques provide to meet the increasing demands of future communication systems. The primary focus was on the employment of MIMO systems including related experimental results from [28, 29]. While the use of antenna arrays is becoming universal, it is indeed the simultaneous utility of arrays at both base station and mobile terminals which can take advantage of multiple signal dimensions and ultimately lead to immense increases in capacity and data rates [28]. To exhibit this argument, simulation results performed for cellular systems and WLANs have been included. These results show the great potentials of building very high speed wireless links.

However, a number of hurdles must be overcome to reap the benefits using MIMO systems. First of all, the proper assessment of the antenna arrangement and spacings and the scattering richness must be guaranteed for the success of these new concepts before they can be widely implemented. For example, the current trend of miniaturizing the size of the cellular phone diminishes the available space necessary for multiple closely spaced antennas.

CHAPTER 9

Commercial Availability of Smart Antennas

Smart antenna systems are designed to relieve capacity strain in cell sites experiencing heavy and imbalanced traffic distribution. Since traffic tends to vary considerably within a network, at an individual cell site and over time, it is critical for wireless service providers to allocate efficiently infrastructure and spectrum resources to meet network capacity and performance demands.

Although smart antennas date back since the late 1950s, it is only because of today's advancement in powerful low-cost digital signal processors, general-purpose processors (and ASICs – Application Specific Integrated Circuits), as well as innovative software-based signal processing techniques (algorithms), that smart antenna systems are gradually becoming commercially available.

Several papers and studies in the area have been published in the recent years and a growing number of companies are studying MIMO systems for future use. Several companies have been created recently that are trying to commercialize MIMO systems and a number of larger companies are also studying this area. In [94] the author well clarifies the main reasons which explain why smart antenna-enabled systems have not yet been deployed more widely. These are repeated here, verbatim, as stated in [94]:

- "DSP and CPU speeds need to be fast enough to handle the increased computations needed for smart antenna algorithms to be able to be implemented in realtime. This was a big problem until 1999 or so, particularly for low-cost solutions. However, computing power has now progressed to the point where smart antenna systems can feasibly be installed using inexpensive processors."

- "The value proposition provided by smart antenna systems needs to outweigh the additional cost of a smart antenna system. In the early days of advanced mobile phone systems, one base station could cover a large area and there was no need for multiplying the capacity of a base station. Today, when microcells and picocells need to be used at times and spectrum is a precious commodity, the differentiation that smart antennas provide justifies the additional cost in a variety of scenarios. As the value of the capacity/range of a system grows to clients and the cost of implementing such systems drops due to continual advances in the field, it is expected that the usage of smart antennas will continue to grow."

- "The number of people that truly understand how smart antennas work is limited. Each year the number of people in this group grows, especially due to funding research in universities and commercial projects, but the supply of experts is limited. A serious problem is the lack of universities offering classes in smart antennas."

- "Decision-makers in the wireless industry have experienced a high level of scepticism about implementing smart antennas, partly due to a lack of understanding on the subject and partly because the systems were not proven to work in commercial environments. The successes have helped to pacify those worries, as have the various test-beds created by academia."

A company that has been able to successfully commercialize smart antenna systems for cellular base stations worldwide is ArrayComm (an innovative wireless technology company located in San Jose, California whose chairman, CEO and co-founder is the inventor of the cellular phone, Martin Cooper). The company has patented the smart antenna system under the name of IntelliCell®, and as of today, it has deployed over 275,000 IntelliCell® base stations in USA, Japan, China, Taiwan, Australia, South Africa, Thailand, Middle East, and Philippines. It is considered a fully adaptive smart antenna technology which dynamically adjusts signal patterns to and from the desired subscriber, creating a concentration of energy focused exclusively on the subscriber for efficient delivery of voice or data services. The technology can concentrate energy on people even as they move, reducing radio interference and giving users the best signal quality possible [45].

ArrayComm's IntelliCell solution uses an array of ordinary antennas to continually map the RF environment. Mapping allows the system to focus on the subscriber, using the environment to coherently combine the signals on the subscriber's device. Less power is transmitted, and less interference is generated, resulting in superior quality of service for the user. A base station utilizing IntelliCell employs a small collection (array) of simple, off-the-shelf antennas (typically 4 to 12) coupled with sophisticated signal processing to manage the energy radiated and received by the base station. This improves coverage and signal quality and mitigates interference in the network on both the uplink and the downlink.

ArrayComm has shown that its IntelliCell® technology has significantly improved capacity and coverage; in fact, on the average, an IntelliCell-enabled network can deliver three times the capacity of a conventional system or up to twice the coverage area, depending on the air interface being deployed.

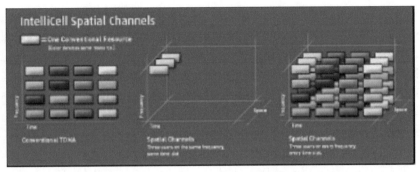

(a)   (b)

FIGURE 9.1: (a) Smart antenna array with 12 elements developed by ArrayComm Inc. and (b) Joint TD-SD Multiple Access scheme by ArrayComm [45].

Moreover, commercialized GSM base stations incorporating ArrayComm technology have proven up to 600% greater capacity than standard GSM networks and frequency reuse of 1 across cells without frequency hopping. The expanded range of IntelliCell-enhanced base stations means fewer cell sites are required resulting in lower costs. It uses low power amplifiers resulting in higher reliability and lower costs.

This enhancement provides enormous potential revenue gains for GSM network operators wanting to maximize their infrastructure and spectrum investments. Company's executives claim that their equipment cost is just 15% higher than conventional equipment for up to three to seven times capacity improvement. Fig. 9.1(a) shows the smart antenna manufactured by ArrayComm Inc.

As previously stated, SDMA is an enhancement to the more common TDMA, FDMA, and CDMA methodologies. Adding a spatial dimension to these accessing schemes creates an additional method of identifying each individual user. This means that instead of a single user being served by one unit of time, frequency or code, that same single-unit can now serve multiple users, identifying each by their unique *spatial signature* [45]. Fig. 9.1(b) illustrates the joint Time Division and Spatial Division Multiple Access scheme realized by the IntelliCell-enabled network. For further details on ArrayComm and its products refer to [45, 227].

Another company that has made smart antenna commercially available is Metawave Communications located in Redmond, Washington. The company's smart antenna offerings provide wireless operators, tower providers and infrastructure manufacturers with cost-effective solutions that maximize capacity and performance, improve quality and increase efficiency of CDMA, GSM and third generation (3G) wireless networks. Metawave's smart antenna solutions have been deployed in 14 of the top 20 markets in the US and five of the nine regions in Mexico.

The company's smart antenna system is patented under the name of Spotlight® and it has installed, as of October 2002, 420 Spotlight® base stations in USA, Central and South America, and Russia. Spotlight® customizes sectors to balance traffic loading. In other words, it is not a fully adaptive smart antenna but a Switched-Beam smart antenna which allows traffic load balance at cell sites and reduction of handoff overhead. Spotlight® provides cellular base stations retro-fitting at a lower cost since it requires less digital signal processing than fully adaptive systems.

Systems with Spotlight® deployment have delivered capacity gains of up to 50% in three-sector sites and over 90% in six-sector sites for CDMA. For further detail on Metawave and its products refer to [227, 228]. Reports from Metawave claim up to a three times capacity increase with adaptive beam-forming. Fig. 9.2 illustrates the functional diagram of Spotlight®.

Another product by the same company is the SmartCell®. It is a targeted, sector-by-sector smart antenna solution for cellular and PCS networks that enables wireless operators to sculpt or shape a cell site's coverage pattern in a way that delivers greater performance, capacity and coverage benefits. The SmartCell system comprises a set of phased-array antenna panels with a customizable "personality module" that establishes an optimally sculpted antenna pattern for a particular sector. A software tool is used to determine the optimal sector antenna that is captured in the personality module and installed into the back of each antenna panel. By

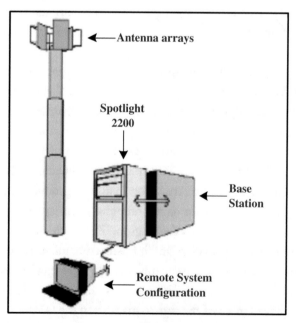

FIGURE 9.2: The functional diagram of Spotlight® [228].

FIGURE 9.3: The use of standard or custom patterns to fit the topology of sector coverage [228].

swapping out the personality module, the embedded *Cell sculpting* technology in the systems provides flexibility to change antenna patterns for determination of the optimal in response to the changing RF environment [228]. This technology is the key to enabling operators to shape a cell's coverage pattern in a way that delivers greater performance, capacity and quality benefits than those of off-the-shelf antennas. Cell sculpting technology takes drive test data and network information to estimate the optimal antenna pattern. The optimal pattern is then transferred to the personality module that is inserted into the back of each antenna panel. Fig. 9.3 illustrates the SmartCell concept.

In complex RF environments without the presence of a dominant server, cell sculpting technology helps increase server dominance in the network to reduce pilot pollution, reduce average transmit power on the forward and reverse link, and reduce variance of transmit power. SmartCell is ideal for complex RF environments where network topology and traffic distributions create difficult radio management challenges. As this changes, operators can make a corresponding change to their antenna patterns. SmartCell technology supports all major air interfaces including CDMA, GSM, TDMA, CDMA2000, and W-CDMA.

Not only smart antennas for cellular systems have come a long way, but also smart antennas for PEDs have made progress. Although their commercial availability has been hindered due to its high costs, research and experiments have shown promising results. A company in Berne, Switzerland, called ASCOM AR&T, has developed a 3-element low power smart antenna for 5–6 GHz W-LAN which is small enough for mobile terminals. The antennas are attached to a PCMCIA card. Each single element is a bent stacked slot antenna which experiences effectively independent fading. Finally, the beamforming is performed at RF frequencies to keep the production costs low. Tests carried out with this product showed a superior performance over an omnidirectional antenna like Bluetooth™ [229]. Other experimental projects in the field

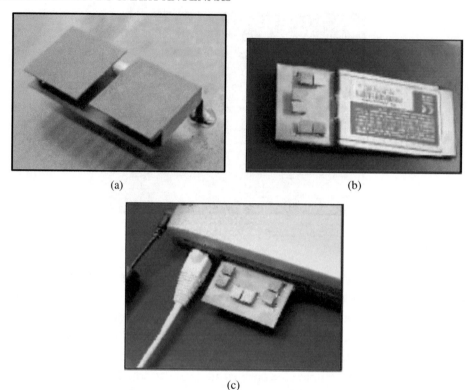

(a)   (b)

(c)

FIGURE 9.4: Smart antenna developed by ASCOM AR&T (a) Single antenna element, (b) PCMCIA card with antenna array, and (c) Smart antenna connected to a Notebook [229].

of smart antennas for W-LANs have also reported similar findings [230, 231]. Fig. 9.4 shows the company's development.

Finally, it should be stated that as the client's demand for higher capacity/range in a system grow and the cost of implementing such systems drops due to continual advances in the field, it is expected that the usage of smart antennas will continue to grow.

CHAPTER 10

Summary

This book examined and analyzed various system aspects of a modern communication system based on smart antenna technology. The analysis began with a presentation of the current communication systems with emphasis on their limitations and challenges that need to be resolved in order to meet the continuous increasing demands of high data rates and capacity of the wireless era.

To better understand the smart antenna technology, an entire chapter was devoted to the properties of antenna elements and arrays, and the classification of antennas according to their radiation characteristics. The major analysis of smart antennas was carried out in the chapters that followed where the functional principles of smart antennas were considered, different smart antenna configurations were suggested and the benefits and drawbacks concerning their commercial introduction were stressed. Smart antenna was then examined from the signal processing point of view. In particular, the fundamental properties of the direction of arrival were detailed and this information was exploited in a way to design the array to appropriately shape its radiation pattern. The subsequent chapter presented the results of an effort to integrate various aspects of smart antenna systems, a project that examined antenna design, adaptive beamforming algorithms and their impact on the communication *channel BER* and network *throughput*.

Afterward, the unique advantages of joint space–time processing techniques were reviewed and its origins and applications were demonstrated. The chapter was also concerned with the attractive characteristics of MIMO systems, including experimental results, a modern technique that exhibits great promise for large data rates and capacities. Lastly, commercial efforts on smart antennas were briefly summarized. Temporal processing has reached very high levels and has become mature, but by itself is not sufficient. However, when combined with space processing, it may be in a position to meet the ever expanding demands of high speed and reliable communication enjoyed by a constantly increasing population. There is no better verification of this argument than the words of Andrew Viterbi, a pioneer in the global spread of wireless communications, *"Spatial processing remains as the most promising, if not the last frontier, in the evolution of multiple access systems"* [232].

Acknowledgments

The authors would like to express their sincere appreciation for the cooperation, suggestions, generous contributions and supply of information by many of the authors of papers from which material in this paper was derived from and based upon. In particular, the authors would like to recognize: Profs. J. R. Mosig and A. Skrivervik, and their graduate student I. Stevanović from Ecole Polytechnique Fédérale de Lausanne, Switzerland; Profs. R. D. Murch and K. B. Letaief from The Hong Kong University of Science and Technology; Mr. J. Baltersee from Aachen University of Technology, Aachen, Germany; Dr. P. H. Lehne and Dr. M. Pettersen from Telenor Research and Development, Fornebu, Norway; Prof. Steven Blostein and his former students, J. Chou and W. Y. Shiu, from Queen's University, Kingston, Ontario, Canada; Prof. Arogyaswami Paulraj from Stanford University and Dr. Constantinos Papadias from Bell Labs, Lucent Technologies; Professor A. Lee Swindlehurst from Brigham Young University; Dr. Stefan Werner from Helsinki University of Technology, Finland; Dr. P. Van Rooyen, Founder and CTO of Zyray Wireless, San Diego, CA; Dr. Reinaldo Valenzuela, Dr. Angel Lozano and Dr. Farrokh R. Farrokhi from the Wireless Communications Research Department at Bell Labs, Lucent Technologies; Profs. B. Ottersten and R. Stridh from The Royal Institute of Technology, Stockholm, Sweden; Prof. G. T. Okamoto from Santa Clara University; George Telecki and Brendan Codey, John Wiley and Sons, Interscience Division; and our colleagues Profs. A. S. Spanias, T. M. Duman, and J. M. Capone and graduate students Dr. S. Bellofiore, J. Foutz, R. Govindarajula, and Dr. I. Bahçeci at Arizona State University. In advance, we apologize for any omissions; they are not intentional.

References

[1] R. D. Murch and K. B. Letaief, "Antenna systems for broadband wireless access," *IEEE Commun. Mag.*, Apr. 2002.

[2] P. N. Fletcher and P. Darwood, "Beamforming for circular and semicircular array antennas for low-cost wireless LAN data communication systems," *IEEE Proc. Microw. Antennas Propagat.*, vol. 145, no. 2, pp. 153–158, Apr. 1998. doi:10.1049/ip-map:19981658

[3] J.-A. Tsai and B. D. Woerner, "Adaptive beamforming of uniform circular arrays UCA for wireless CDMA system," in *Record of the Thirty-Fifth Asilomar Conference on Volume Signals, Systems and Computers*, vol. 1, Nov. 2001, pp. 399–403. doi:full_text

[4] "Smart antenna systems," International Engineering Consortium. [Online]. Available: www.iec.org/online/tutorials/smart_antennas

[5] "Smart antennas," CDMA Development Group, 2004. [Online]. Available: http://www.cdg.org/technology/cdma_technology/smart_antennas/index.asp

[6] M. Chryssomallis, "Smart antennas," *IEEE Antennas Propagat. Mag.*, vol. 42, no. 3, pp. 129–136, June 2000. doi:10.1109/74.848965

[7] S. Andersson, M. Millnert, M. Viberg, and B. Wahlberg, "A study of adaptive arrays for mobile communication systems," in *IEEE International Conference on Acoustics, Speech, and Signal Processing*, vol. 5, 1991, pp. 3289–3292.

[8] G. V. Tsoulos, G. E. Athanasiadou, M. A. Beach, and S. C. Swales, *Adaptive Antennas for Microcellular and Mixed Cell Environments with DS-CDMA*. Netherlands, Kluwer, 1998, ch. 7, Wireless Personal Communications, pp. 147–169.

[9] R. Kohno, "Spatial and temporal communication theory using adaptive antenna array," *IEEE Personal Commun. Mag.*, vol. 51, Feb. 1998.

[10] T. S. Rappaport, *Smart Antennas Adaptive Arrays, Algorithms, Wireless Position Locations Selected Readings*. Piscataway, NJ: IEEE, 1998.

[11] G. V. Tsoulos, *Adaptive Antennas for Wireless Communications*. Piscataway, NJ: IEEE, 2001.

[12] A. O. Boukalov and S. G. Häggman, "System aspects of smart-antenna technology in cellular wireless communications – an overview," *IEEE Trans. Microw. Theory Tech.*, vol. 48, no. 6, pp. 919–929, June 2000. doi:10.1109/22.846718

[13] J. C. Liberti and T. S. Rappaport, *Smart Antennas for Wireless Communications: IS-95 and Third Generation CDMA Applications.*   Upper Saddle River, NJ: Prentice Hall PTR, 1999.

[14] G. V. Tsoulos, "Smart antennas for mobile communication systems; benefits and challenges, *IEEE Commun. Eng. J.,* vol. 11, no. 2, pp. 84–94, Apr. 1999.

[15] W.-S. Wang, "Bluetooth: A new era of connectivity," *IEEE Microw. Mag.,* vol. 3, no. 3, pp. 38–42, Sept. 2002. doi:10.1109/MMW.2002.1028360

[16] M. Bravo-Escos, "Networking gets personal," *IEE Rev.,* vol. 48, no. 1, pp. 32–36, Jan. 2002. doi:10.1049/ir:20020104

[17] S. Bellofiore, "Smart antenna systems for mobile platforms," Ph.D. dissertation, Arizona State University, Dec. 2002.

[18] E. G. Larsson, P. Stoica, and G. Ganesan, *Space–time Block Coding for Wireless Communications.* Cambridge: Cambridge University Press, June 2003.

[19] J. Y.-L. Chou, "An investigation on the impact of antenna array geometry on beamforming user capacity," Master's thesis, Queen's University, Kingston, Ontario, Mar. 2002.

[20] I. Stevanović, A. Skrivervik, and J. R. Mosig, "Smart antenna systems for mobile communications," Ecole Polytechnique Fédérale de Lausanne, Lausanne, Suisse, Tech. Rep., Jan. 2003. [Online]. Available: http://lemawww.epfl.ch

[21] A. Paulraj, B. Ottersten, R. Roy, A. Swindlehurst, G. Xu, and T. Kailath, *Subspace Methods for Direction of Arrival Estimation.* Amsterdam: North-Holland, 1993, vol. 10, ch. 16, pp. 693–739.

[22] W. Y. Shiu, "Noniterative digital beamforming in CDMA cellular communications systems," Master's thesis, Queen's University, Kingston, Ontario, Nov. 1998.

[23] S. Werner, "Reduced complexity adaptive filtering algorithms with applications to communications systems," Ph.D. dissertation, Helsinki University of Technology, Helsinki, Finland, Oct. 2002.

[24] S. Bellofiore, J. Foutz, J. Govindarajula, İsrafil Bahçeci, C. A. Balanis, A. S. Spanias, J. M. Capone, and T. M. Duman, "Smart antenna system analysis, integration and performance for mobile ad-hoc network (MANETs)," *IEEE Trans. Antennas Propagat.,* vol. 50, no. 5, p. 571–581, May 2002.

[25] P. Van Rooyen, "Advances in space–time processing techniques open up mobile apps," Nov. 2002. [Online]. Available: http://www.eetimes.com/in_focus/mixed_signals/OEG20021107S0021

[26] A. J. Paulraj, D. Gesbert, and C. Papadias, *Encyclopedia for Electrical Engineering.* New York: Wiley, 2000, ch. Antenna arrays for Wireless Communications, pp. 531–563.

[27] A. J. Paulraj and C. B. Papadias, "Space–time processing for wireless communications," *IEEE Signal Process. Mag.*, vol. 14, no. 6, pp. 49–83, Nov. 1997. doi:10.1109/79.637317

[28] A. Lozano, F. R. Farrokhi, and R. A. Valenzuela, "Lifting the limits on high speed wireless data access using antenna arrays," *IEEE Commun. Mag.*, vol. 39, no. 9, pp. 156–162, Sept. 2001. doi:10.1109/35.948420

[29] R. Stridh and B. Ottersten, "Spatial characterization of indoor radio channel measurements at 5 GHz," in *First IEEE Sensor Array and Multichannel Signal Procesing Workshop*, Cambridge, Massachusetts, USA, Mar. 2000. [Online]. Available: http://www.s3.kth.se/radio/4GW/public/Papers/RickardStridhSAM2000.pdf

[30] International Telecommunications Union (ITU). [Online]. Available: http://www.itu.int/home

[31] http://www.wirelessintelligence.com

[32] Alexander Resources. [Online]. Available: http://www.alexanderresources.com/reports/Report2/Summary.htm

[33] Micrologic Research. [Online]. Available: http://www.mosmicro.com

[34] Micrologic Research. [Online]. Available: http://www.mosmicro.com/Cell_Exec.pdf

[35] International Engineering Consortium. [Online]. Available: http://www.iec.org/online/tutorials/gsm/topic01.html

[36] Qualcomm Corporation. [Online]. Available: http://www.qualcomm.com/cdma

[37] Ericsson. [Online]. Available: http://www.ericsson.com/cdmasystems/3gcdma2000.html

[38] "Air Interface Fundamentals: UMTS and W-CDMA," Award Solutions. [Online]. Available: http://www.awardsolutions.com/training/elearning_pdfs

[39] T. Rappaport, "The wireless communications revolution: Past, present, & future," Virginia Tech, Tech. Rep., Aug. 1997. [Online]. Available: http://www.mprg.org/Tech_xfer/ppt/vt125th.pdf

[40] L. C. Godara, "Applications of antenna arrays to mobile communications. Part I: Performance improvement, feasibility, and system considerations," in *Proc. IEEE*, vol. 85, pp. 1031–1060, July 1997. doi:10.1109/5.611108

[41] P. M. Shankar, *Introduction to Wireless Systems.*   New York: Wiley, 2002.

[42] M. C. Vanderveen, "Estimation of parametric channel models in wireless communications networks," Ph.D. dissertation, Stanford University, Department of Scientific Computing and Computational Mathematics, Nov. 1997.

[43] J. Baltersee, "Smart Antennas and Space-Time Processing," Aachen University of Technology, Institute for Integrated Signal Processing Systems, Tech. Rep., May 1998.

[44] V. K. Garg and J. E. Wilkes, *Wireless and Personal Communications Systems.* Upper Saddle River, NJ: Prentice Hall PTR, 1996.

[45] "Intellicell: Bringing Wireless to Life," Arraycomm, 2003. [Online]. Available: http://www.arraycomm.com

[46] B. Pattan, *Robust Modulation Methods & Smart Antennas in Wireless Communications*. Upper Saddle River, NJ: Prentice Hall PTR, 2000.

[47] T. I. Song, D. J. Kim, and C. H. Cheon, "Optimization of sectorized antenna beam patterns for CDMA2000 systems." in *3G Mobile Communications Technologies, Conference Publication, No. 471 Third International Conference on 3G Mobile Communication Technologies, 2002. (Conf. Publ. No. 489)*, May 2002, pp. 428–432.

[48] X. Yang, S. Ghaheri, R. Niri, and R. Tafazzoli, "Sectorization gain in CDMA cellular systems." in *Third International Conference on 3G Mobile Communication Technologies, 2002. (Conf. Publ. No. 489)*, 2000, pp. 70–75.

[49] T. M. Cover and J. A. Thomas, *Elements of Information Theory*. New York: Wiley-Interscience, Aug. 1991.

[50] D. Cox, H. Arnold, and P. Porter, "Universal digital portable communications: A system perspective," *IEEE J. Select. Areas Commun.*, vol. 5, no. 5, pp. 764–773, June 1987. doi:10.1109/JSAC.1987.1146603

[51] S. Hara and R. Prasad, "Overview of multicarrier CDMA," *IEEE Commun. Mag.*, vol. 35, pp. 126–133, Dec. 1997. doi:10.1109/35.642841

[52] D. Huff, "Direct Sequence Spread Spectrum Tutorial." [Online]. Available: www.eas.asu.edu/$\sim$apapand/applet/TF_JAVA/applet/DSSS_tutorial.doc

[53] W. C. Y. Lee, "Overview of cellular CDMA," *IEEE Trans. Veh. Technol.*, vol. 40, pp. 291–302, 1991. doi:10.1109/25.289410

[54] M. Engels, *Wireless OFDM Systems: How to Make Them Work?*, M. Engels, Ed. Dordrecht Kluwer, 2002.

[55] [Online]. Available: http://www.wireless.per.nl:202/telelearn/ofdm/

[56] Z. Li, "Performance of multicarrier ds-cdma systems using mutually orthogonal complementary sets of sequences," University of Florida, Tech. Rep., 2000. [Online]. Available: http://www.tec.ufl.edu/$\sim$zbli/report.pdf

[57] M. Cooper and M. Goldburg, "Intelligent antennas: Spatial division multiple access." Wireless, *Anual. Revw. of Commun.*, pp. 999–1002, 1996, ArrayComm Inc., San Jose, CA.

[58] C. A. Balanis, "Antenna theory: A review," in *Proc. IEEE*, vol. 80, no. 1, Jan. 1992, pp. 7–23. doi:10.1109/5.119564

[59] C. A. Balanis, *Antenna Theory: Analysis and Design*, 3rd ed. New York: Wiley, 2005.

[60] P.-J. Wan, "Capacity expansion: Sectorized cellular systems." [Online]. Available: http://www.csam.iit.edu/$\sim$wan/lecture05.pdf

[61] R. T. Compton Jr., R. Huff, W. G. Swarner, and A. A. Ksienski, "Adaptive arrays for communication systems: An overview of research at the ohio state university," *IEEE Trans. Antennas Propagat.*, vol. AP-24, no. 5, pp. 599–607, Sept. 1976. doi:10.1109/TAP.1976.1141413

[62] R. A. Monzingo and T. W. Miller, *Introduction to Adaptive Arrays.* Scitech Publishing Inc., Oct. 2003, Mendham, NJ. ISBN 1-891121-24-3.

[63] I. Gupta and A. Ksienski, "Dependence of adaptive array performance on conventional array design," *IEEE Trans. Antennas Propagat.*, vol. 30, no. 4, pp. 549–553, July 1982. doi:10.1109/TAP.1982.1142867

[64] A. Ishide and R. T. Compton Jr., "On grating nulls in adaptive arrays," *IEEE Trans. Antennas Propagat.*, vol. 28, no. 4, pp. 467–475, July 1980.

[65] S. P. Applebaum, "Adaptive arrays," *IEEE Trans. Antennas Propagat.*, vol. AP-24, no. 5, pp. 585–598, Sept. 1976. doi:10.1109/TAP.1976.1141417

[66] Z. Fu, "Adaptive arrays antenna systems," Ithaca, NY 14850, USA. [Online]. Available: http://people.cornell.edu/pages/zf24/Adaptive_arrays.htm

[67] D. G. Brennan, "Linear diversity combining techniques," in *Proc. IRE*, vol. 47, 1959, pp. 1075–1102. doi:10.1109/JRPROC.1959.287136

[68] G. L. Stüber, *Principles of Mobile Communication.* Dodrecht: Kluwer, 2001.

[69] S. W. Kim, D. S. Ha, and J. H. Kim, "Performance gain of smart dual antennas at handsets in 3G CDMA system," in *CDMA International Conference*, vol. 2, Nov. 2000, pp. 223–227.

[70] İsrafil Bahçeci, "Trellis- and turbo-coded modulation for multiple antennas over fading channels," Master's thesis, Arizona State University, Aug. 2001.

[71] J. G. Proakis, *Digital Communications*, 4th ed. New York: McGraw-Hill, 2001.

[72] W. C. Jakes, *Microw. Mobile Communications.* New York: Wiley, 1974.

[73] S. Glisic and B. Vucetic, *Spread Spectrum CDMA Systems for Wireless Communications.* Boston, MA: Artech House, 1997.

[74] Z. Zhao, S. Stapleton, and J. K. Cavers, "Analysis of polarization diversity scheme with channel codes," Department of Engineering Science, Simon Fraser University, Burnaby, BC, Canada, V5A 1S6, Tech. Rep., 1999.

[75] [Online]. Available: http://www.wirelessreview.com/ar/wireless_space_vs_polarization

[76] "Application of Angular Diversity Systems and Discussion of Antenna Pattern Measurements." [Online]. Available: http://www.nsma.org/recommendation/wg16-89-06.pdf

[77] P. L. Perini and C. L. Holloway, "Angle and space diversity comparisons in different mobile radio environments," *IEEE Trans. Antennas Propagat.*, vol. 46, no. 6, pp. 957–1000, June 1998. doi:10.1109/8.686760

[78] "Special issue on adaptive antennas," *IEEE Trans. Antennas Propagat.*, vol. 24, no. 5, Sept. 1976.

[79] "Special issue on adaptive processing antenna systems," *IEEE Trans. Antennas Propagat.*, vol. 34, no. 3, Mar. 1986.

[80] P. H. Lehne and M. Pettersen, "An overview of smart antenna technology for mobile communications systems," *IEEE Communications Surveys*, vol. 2, no. 4, pp. 2–13, Fall Quarter 1999. [Online]. Available:
www.comsoc.org/livepubs/surveys/public/4q99issue/pdf/Lehne.pdf

[81] J. H. Winters, "Smart antennas for wireless systems," *IEEE Personal Commun. Mag.*, vol. 5, no. 1, pp. 23–27, Feb. 1998. doi:10.1109/98.656155

[82] A. Scherb, V. Kühn, and K.-D. Kammeyer, "Comparison of intelligent code acquisition for sectorized multi-antenna CDMA in downlink mode," University of Bremen, Department of Communications Engineering, Otto-Hahn-Allee, D-28359 Bremen, Germany, Tech. Rep., 2003. [Online]. Available:
http://www.ant.uni-bremen.de/whomes/scherb/puplications/1180757472.pdf

[83] D. H. Johnson and D. E. Dudgeon, *Array Signal Processing: Concepts and Techniques*. Englewood Cliffs, NJ: Prentice-Hall, 1992.

[84] A. R. Lopez, "Performance predictions for cellular switched beam inelligent antnna systems," *IEEE Commun. Mag.*, vol. 34, no. 10, pp. 152–154, Oct. 1996. doi:10.1109/35.544336

[85] T. Thrassyvoulou, "Adaptive beamforming: Using a complex bounding ellipsoid algorithm with gradient projections," Master's thesis, Arizona State University, Aug. 2003.

[86] J. H. Winters, WTEC Panel Report on Wireless Technologies and Information Networks, Chapter 6, Smart Antennas. International Technology Research Institute, Baltimore, MD, July 2000.

[87] J. Butler and R. Lowe, "Beam-forming matrix simplifies design of electronically scanned antennas," *Electron. Des.*, vol. 9, no. 8, pp. 1730–1733, Apr. 1961.

[88] Y. T. Lo and S. W. Lee, *Antenna Handbook*. New York: Van Nostrand Reinhold Company Inc., 1988.

[89] C. B. Dietrich, Jr., "Adaptive arrays and diversity antenna configurations for handheld wireless communication terminals," Ph.D. dissertation, Virginia Polytechnic Institute and State University, Blacksburg, VA, Feb. 2000.

[90] N. C. T. Desmond, "Smart antennas for wireless applications and switched beamforming," The University of Queensland, Brisbane, Australia, Tech. Rep., Oct. 2001.

[91] J. R. James and P. S. Hall, *Handbook of Microstrip Antennas*. London: Peregrinus on behalf of Institution of Electrical Engineers, 1989.

[92] T. S. N. Chan, "Butler matrix feed configuration for phased array," University of Queensland, Department of Electrical and Computer Engineering, Brisbane, Australia, Tech. Rep., Oct. 1994.

[93] T. Do-Hong and P. Russer, "Signal processing for wideband array applications," *IEEE Microw.*, vol. 5, no. 1, pp. 57–67, Mar. 2004. doi:10.1109/MMW.2004.1284944

[94] G. Okamoto, "Developments and advances in smart antennas for wireless communications," Santa Clara University, Tech. Rep., 2003. [Online]. Available: www.wmrc.com/businessbriefing/pdf/wireless_2003/Publication/okamoto.pdf

[95] F. Shad, T. D. Todd, V. Kezys, and J. Livta, "Dynamic slot allocation (dsa) in indoor sdma/tdma using a smart antenna basestation," *IEEE/ACM Trans. Networking*, vol. 9, no. 1, pp. 69–81, Feb. 2001. doi:10.1109/90.909025

[96] C. Ung and R. H. Johnston, "A space division multiple access receiver," in *IEEE International Symposium on Antennas and Propagation*, vol. 1, pp. 422–425, 2001.

[97] M. Ghavami and R. Kohno, "A new broadband uniform accuracy DOA estimator," *Accepted for publication in the European Transactions on Telecommunications*, 2002. [Online]. Available: http://www.csl.sony.co.jp/ATL/papers/ghp3.pdf

[98] J. Litva and T. Lo, *Digital Beamforming in Wireless Communications.* Boston, MA: Artech House Publishers, 1996.

[99] V. Viopio, "Adaptive antennas," HUT Radio Laboratory, Tech. Rep., 1998.

[100] J. S. Thompson, P. M. Grant, and B. Mulgrew, "Smart antenna arrays for CDMA systems," *IEEE Pers. Commun. Mag.*, vol. 3, no. 5, pp. 16–25, Oct. 1996. doi:10.1109/98.542234

[101] R. G. Vaughan, "On optimum combining at the mobile," *IEEE Trans. Veh. Technol.*, vol. 37, pp. 181–188, Nov. 1988. doi:10.1109/25.31122

[102] S. C. K. Ko and R. D. Murch, "On optimum combining at the mobile," *IEEE Trans. Antennas Propagat.*, vol. 49, pp. 954–960, June 2001. doi:10.1109/8.931154

[103] J. S. Colburn, Y. Rahmat-Samii, and M. A. Jensen, "Diversity performance of dual antenna personal communication handsets," in *Proc. IEEE Antennas Propagat. Soc. Int. Symp. Dig.*, July 1996, pp. 730–733. doi:full_text

[104] T. A. Denidni, D. McNeil, and G. Y. Delisle, "Experimental investigations of a new adaptive dual-antenna array for handset applications," *IEEE Trans. Veh. Technol.*, vol. 52, pp. 1417–1423, Nov. 2003. doi:10.1109/TVT.2003.816646

[105] L. C. Godara, "Applications of antenna arrays to mobile communications. Part II: Beam-forming and direction-of-arrival considerations," in *Proc. IEEE*, vol. 85, Aug. 1997, pp. 1195–1245. doi:10.1109/5.622504

[106] L. C. Godara and A. Cantoni, "Uniqueness and linear independence of steering vectors in array space," *J. Acoust. Soc. Am.*, vol. 70, pp. 467–475, 1981. doi:10.1121/1.386790

[107] [Online]. Available: http://mathworld.wolfram.com

[108] I. J. Gupta and A. A. Ksienski, "Effect of mutual coupling on the performance of adaptive arrays," *IEEE Trans. Antennas Propagat.*, vol. AP-31, no. 5, pp. 785–791, Sept. 1983. doi:10.1109/TAP.1983.1143128

[109] H. Steyskal and J. S. Herd, "Mutual coupling compensation in small array antennas," *IEEE Trans. Antennas Propagat.*, vol. 38, no. 12, pp. 1971–1975, Dec. 1990. doi:10.1109/8.60990

[110] T. Svantesson, "Direction finding in the presence of mutual coupling," Thesis for the degree of Licentiate of Engineering, Chalmers University of Technology, School of Electrical and Computer Engineering, Department of Signals and Systems, Göteborg, Sweden, Tech. Rep., 1999.

[111] T. Su and H. Ling, "On modeling mutual coupling in antenna arrays using coupling matrix," *Microw. Opt. Technol. Lett.*, vol. 28, no. 4, pp. 231–237, Feb. 2001. doi:10.1002/1098-2760(20010220)28:4<231::AID-MOP1004>3.0.CO;2-P

[112] H. T. Hui, "Improved compensation for the mutual coupling effect in a dipole array for direction finding," *IEEE Trans. Antennas Propagat.*, vol. 51, no. 9, pp. 2498–2503, Sept. 2003. doi:10.1109/TAP.2003.816303

[113] Z. Huang, C. A. Balanis, and C. R. Birtcher, "Mutual coupling compensation in UCAs: Simulations and experiment," *IEEE Trans. Antennas Propagat.*, vol. 54, no. 11, pp. 3082–3086, Nov. 2006. doi:10.1109/TAP.2006.883989

[114] S. M. Kay, *Fundamentals of Statistical Signal Processing, Volume I: Estimation Theory*. Prentice Hall PTR, Upper Saddle River, NJ, Mar. 1993.

[115] H. Messer, Y. Rockah, and P. M. Schultheiss, "Localization in the presence of coherent interference," *IEEE Trans. Acoust., Speech, Signal Process.*, vol. 38, no. 12, pp. 2025–2032, Dec. 1990. doi:10.1109/29.61530

[116] A. N. Mirkin and L. H. Sibul, "Cramer-Rao bounds on angle estimation with a two-dimensional array," *IEEE Trans. Signal Process.*, vol. 39, no. 2, pp. 515–517, Feb. 1991. doi:10.1109/78.80843

[117] R. O. Nielsen, "Estimation of azimuth and elevation angles for a plane wave sine wave with a 3-D array," *IEEE Trans. Signal Process.*, vol. 42, no. 11, pp. 3274–3276, Nov. 1994. doi:10.1109/78.330396

[118] J. Goldberg and H. Messer, "Inherent limitations in the localization of a coherently scattered source," *IEEE Trans. Signal Process.*, vol. 46, no. 12, pp. 3441–3444, Dec. 1998. doi:10.1109/78.735321

[119] A. Dogandzic and A. Nehorai, "Cramer-Rao bounds for estimating range, velocity, and direction with an active array," *IEEE Trans. Signal Process.*, vol. 49, no. 6, pp. 1122–1137, June 2001. doi:10.1109/78.923295

[120] W. P. Ballance and A. G. Jaffer, "The explicit analytic Cramer-Rao bound on angle estimation," in *22nd Asilomar Conference on Signals, Systems and Computers*, vol. 1, Oct. 31–Nov. 2 1988, pp. 345–351. doi:full_text

[121] A. Bhuyan and P. M. Schultheiss, "Estimation of source separation with an array of arbitrary shape," in *International Conference on Acoustics, Speech, and Signal Processing*, vol. 5, 3–6 April 1990, pp. 2771–2774.

[122] R. Roy and T. Kailath, "ESPRIT-estimation of signal parameters via rotational invariance techniques," *IEEE Trans. Acoust. Speech Signal Process.*, vol. 37, no. 7, pp. 984–995, July 1989. doi:10.1109/29.32276

[123] R. O. Schmidt, "A signal subspace approach to multiple emitter location and spectral estimation," Ph.D. dissertation, Stanford University, 1981.

[124] A. L. Swindlehurst and T. Kailath, "Azimuth/elevation direction finding using regular array geometries," *IEEE Trans. Aerosp. Electron. Syst.*, vol. 29, no. 1, pp. 145–156, Jan. 1993. doi:10.1109/7.249120

[125] C. Chambers, T. C. Tozer, K. C. Sharman, and T. S. Durrani, "Temporal and spatial sampling influence on the estimates of superimposed narrowband signals: when less can mean more," *IEEE Trans. Acoust. Speech Signal Process.*, vol. 44, no. 12, pp. 3085–3098, Dec. 1996.

[126] R. Schmidt, "Multiple emitter location and signal parameter estimation," *IEEE Trans. Antennas Propagat.*, vol. 34, no. 3, pp. 276–280, Mar. 1986. doi:10.1109/TAP.1986.1143830

[127] A. Swindlehurst, "Alternative algorithm for maximum likelihood DOA estimation and detection," *IEE Proc. Radar Sonar Navig.*, vol. 141, no. 6, pp. 293–299, Dec. 1994. doi:10.1049/ip-rsn:19941366

[128] C. P. Mathews and M. D. Zoltowski, "Eigen-structure techniques for 2-D angle of arrival with uniform circular arrays," *IEEE Trans. Signal Process.*, vol. 42, no. 9, pp. 2395–2407, Sept. 1994. doi:10.1109/78.317861

[129] A. Papoulis and S. U. Pillai, *Probability, Random Variables, and Stochastic Processes*, 4th ed. New York: McGraw-Hill, Dec. 2001.

[130] A. L. Swindlehurst and T. Kailath, "A performance analysis of subspace-based methods in the presence of model errors. Part I: The MUSIC algorithm," *IEEE Trans. Signal Process.*, vol. 40, no. 7, pp. 1758–1774, July 1992. doi:10.1109/78.143447

[131] M. S. Bartlett, "Smoothing periodograms from time series with continuous spectra," Nature, vol. 161, pp. 686–687, 1948.

[132] P. Stoica and R. Moses, *Introduction to Spectral Analysis*. Upper Saddle River, NJ: Prentice-Hall, 1997.

[133] T. Svantesson, "Antennas and propagation from a signal processing perspective," Ph.D. dissertation, Chalmers University of Technology, School of Electrical and Computer Engineering, Department of Signals and Systems, Göteborg, Sweden, June 2001.

[134] S. V. Schell and W. A. Gardner, *Handbook of Statistics*.   Amsterdam: North-Holland, 1993, vol. 10, ch. 18, pp. 755–817.

[135] G. Bienvenu and L. Kopp, "Principle de la goniometrie passive adaptive," in *Proc. 7' eme Colloque GRESIT*, Nice, France, 1979, pp. 106/1–106/10.

[136] A. J. Barabell, J. Capon, D. F. Delong, J. R. Johnson, and K. Senne, "Performance Comparison of Superresolution Array Processing Algorithms," Lincoln Laboratory, M.I.T., Tech. Rep. TST-72, 1984.

[137] R. H. Roy, "ESPRIT—estimation of signal parameters via rotational invariance techniques," Ph.D. dissertation, Stanford University, 1987.

[138] A. L. Swindlehurst and T. Kailath, "A performance analysis of subspace-based methods in the presence of model errors. Part II: Multidimensional algorithms," *IEEE Trans. Signal Process.*, vol. 41, no. 9, pp. 2882–2890, Sept. 1993. doi:10.1109/78.236510

[139] L. J. Gleser, "Estimation in a multivariate "errors in variables" regression model: Large sample results," *Ann. of Stat.*, vol. 9, no. 1, pp. 24–44, 1981.

[140] A. Paulraj, R. Roy, and T. Kailath, "Estimation of signal parameters via rotational invariance techniques[1]-ESPRIT," in *19th Asilomar Conference on Circuits, Systems and Computers*, San Jose, CA, Nov. 1985, pp. 83–89. doi:full_text

[141] R. Roy, A. Paulraj, and T. Kailath, "ESPRIT-a subspace rotation approach to estimation of parameters of cisoids in noise," *IEEE Trans. Acoust. Speech Signal Process.*, vol. 34, no. 5, pp. 1340–1342, Oct. 1986. doi:10.1109/TASSP.1986.1164935

[142] A. Swindlehurst, "DOA identifiability for rotationally invariant arrays," *IEEE Trans. Signal Process.*, vol. 40 , Issue: , July, no. 7, pp. 1825–1828, July 1992. doi:10.1109/78.143455

[143] M. Wax and I. Ziskind, "On unique localization of multiple sources by passive sensor arrays," *IEEE Trans. Signal Process.*, vol. 37, no. 7, pp. 996–1000, July 1989. doi:10.1109/29.32277

[144] B. D. Van Veen and K. M. Buckley, "Beamforming: A versatile approach to spatial filtering," *IEEE ASSP Mag.*, vol. 5, pp. 4–24, Apr. 1988. doi:10.1109/53.665

[145] B. Widrow, P. E. Mantey, L. J. Griffiths, and B. B. Goode, "Adaptive antenna systems," *Proc. IEEE*, vol. 55, no. 12, pp. 2143–2159, Aug. 1967.

[146] R. Gooch, B. Sublett, and R. Lonski, "Adaptive beamformers in communications and direction finding systems," in *24th Asilomar Conference on Signals, Systems and Computers*, vol. 1, 1990, pp. 11–15.

[147] R. T. Compton, *Adaptive Antennas: Concepts and Performance*. Upper Saddle River, NJ: Prentice Hall PTR, Jan. 1988.

[148] M. H. Hayes, *Statistical Digital Signal Prcessing and Modelling*. New York: Wiley, 1996.

[149] O. L. Frost, "An algorithm for linearly constrained adaptive array processing," *Proc. IEEE*, vol. 60, no. 8, pp. 926–935, Aug. 1972.

[150] S. Haykin, *Adaptive Filter Theory*. Englewood Cliffs, NJ: Prentice Hall PTR, 1996.

[151] P. S. R. Diniz, *Adaptive Filtering: Algorithms and Practical Implementations*. Boston, MA: Kluwer, 1997.

[152] G. H. Golub and C. F. V. Loan, *Matrix Computations*, 3rd ed. Baltimore, MO: Johns Hopkins University Press, Nov. 1996.

[153] J. Nagumo and A. Noda, "A learning method for system identification," *IEEE Trans. Automat. Contr.*, vol. 12, no. 3, pp. 282–287, June 1967. doi:10.1109/TAC.1967.1098599

[154] D. T. M. Slock, "On the convergence behavior of the LMS and the normalized LMS algorithms," *IEEE Trans. Signal Process.*, vol. 41, no. 9, pp. 2811–2825, Sept. 1993. doi:10.1109/78.236504

[155] G. C. Goodwin and S. K. Sin, *Adaptive Filtering Prediction and Control*. Englewood Cliffs, NJ: Prentice Hall PTR, 1984.

[156] Z. Rong, "Simulation of adaptive array algorithms for CDMA systems," Master's thesis, Virginia Polytechnic Institute and State University, Blacksburg, VA, Sept. 1996.

[157] B. Widrow and S. D. Stearns, *Adaptive Signal Process.*. Englewood Cliffs, NJ: Prentice Hall PTR, 1985.

[158] S. M. Kay, *Fundamentals of Statistical Signal Process., Volume II: Detection theory*. Prentice Hall PTR, Upper Saddle River, NJ, Jan. 1998.

[159] J. H. Reed, *Software Radio: A Modern Approach to Radio Engineering*. Upper Saddle River, NJ: Prentice Hall PTR, 2002.

[160] K. Phillips, Z. Hu, K. Blankenship, Z. Siddiqi, and N. Correal, "Implementation of an adaptive antenna array using the TMS320C541," Texas Instruments, Tech. Rep., Apr. 1999.

[161] J. Treichler and B. Agee, "A new approach to multipath correction of constant modulus signals," *IEEE Trans. Acoust., Speech, Signal Process.*, vol. 31, no. 2, pp. 459–472, Apr. 1983. doi:10.1109/TASSP.1983.1164062

[162] J. Apolinário Jr., M. L. R. Campos, and P. S. R. Diniz, "Convergence analysis of the binormalized data-reusing LMS algorithm," *IEEE Trans. Signal Process.*, vol. 48, no. 11, pp. 3235–3242, Nov. 2000. doi:10.1109/78.875480

[163]   B. A. Schnaufer, "Practical techniques for rapid and reliable real-time adaptive filtering," Ph.D. dissertation, University of Illinois at Urbana-Champaign, Urbana-Champaign, IL, USA, 1995.

[164]   S. L. Gay and S. Tavathia, "The fast affine projection algorithm," in *International Conference on Acoustics, Speech, and Signal Process.*, vol. 5, May 1995, pp. 3023–3026.

[165]   S. G. Sankaran and A. A. L. Beex, "Convergence behavior of affine projection algorithms," *IEEE Trans. Signal Process.*, vol. 48, no. 4, pp. 1086–1096, Apr. 2000. doi:10.1109/78.827542

[166]   D. T. M. Stock, "The block underdetermined covariance (BUC) fast transversal filter (FTF) algorithm for adaptive filtering," in *26th Asilomar Conference on Signals, Systems and Computers*, vol. 1, Oct. 1992, pp. 550–554.

[167]   M. L. R. de Campos, J. Apolinário Jr., and P. S. R. Diniz, "On normalized data-reusing and affine projection algorithms," in *6th IEEE International Conference on Electronics, Circuits and Systems, ICECS '99*, vol. 2, Pafos, Cyprus, Sept. 1999, pp. 843–846. doi:full_text

[168]   M. L. R. de Campos and A. Antoniou, "A new quasi-Newton adaptive filtering algorithm," *IEEE Trans. Circuits Syst. II*, vol. 44, no. 11, pp. 924–934, Nov. 1997. doi:10.1109/82.644046

[169]   M. L. R. de Campos, "Development and analysis of fast and robust Newton-type adaptation algorithms," Ph.D. dissertation, University of Victoria, British Columbia, Canada, 1995.

[170]   M. L. R. de Campos and A. Antoniou, "Analysis of a quasi-Newton adaptive filtering algorithm," in *3rd IEEE International Conference on Electronics, Circuits and Systems, ICECS '96*, vol. 2, Rhodes, Greece, Oct. 1996, pp. 924–934.

[171]   S. Bellofiore, C. A. Balanis, J. Foutz, and A. S. Spanias, "Smart-antenna systems for mobile communication networks. Part 1: Overview and antenna design," *IEEE Antennas Propagat. Mag.*, vol. 44, no. 3, pp. 145–154, June 2002. doi:10.1109/MAP.2002.1039395

[172]   S. Bellofiore, C. A. Balanis, J. Foutz, and A. S. Spanias, "Smart-antenna systems for mobile communication networks. Part 2: Beamforming and network throughput," *IEEE Antennas Propagat. Mag.*, vol. 44, no. 4, pp. 106–114, Aug. 2002. doi:10.1109/MAP.2002.1043158

[173]   S. Bellofiore, C. A. Balanis, J. Foutz, and A. S. Spanias, "Smart antennas for wireless communications," in *IEEE Antennas and Propagation Society International Symposium*, vol. 4, July 2001, pp. 26–29.

[174] S. Bellofiore, C. A. Balanis, J. Foutz, and A. S. Spanias, "Impact of smart antenna designs on network capacity," in *IEEE Antennas and Propagation Society International Symposium*, vol. 3, June 2002, pp. 210–213.

[175] IEEE Std. 802.11, Nov. 1997, IEEE Standard for Wireless LAN Medium Access Control (MAC) Physical Layer (PHY) Specifications.

[176] M. D. Zoltowski, M. Haardt, and C. P. Mathews, "Closed-form 2-D angle estimation with rectangular arrays in element space or beamspace via unitary ESPRIT," *IEEE Trans. Signal Process.*, vol. 44, no. 2, pp. 316–328, Feb. 1996. doi:10.1109/78.485927

[177] P. Strobach, "Two-dimensional equirotational stack subspace fitting with an application to uniform rectangular arrays and ESPRIT," *IEEE Trans. Signal Process.*, vol. 48, no. 7, pp. 1902–1914, July 2000. doi:10.1109/78.847777

[178] A. L. Swindlehurst, B. Ottersten, R. Roy, and T. Kailath, "Multiple invariance ESPRIT," *IEEE Trans. Signal Process.*, vol. 40, no. 4, pp. 867–881, Apr. 1992. doi:10.1109/78.127959

[179] M. Viberg and B. Ottersten, "Sensor array processing based on subspace fitting," *IEEE Trans. Signal Process.*, vol. 39, no. 5, pp. 1110–1121, May 1991. doi:10.1109/78.80966

[180] P. Strobach, "Bi-iteration multiple invariance subspace tracking and adaptive ES-PRIT," *IEEE Trans. Signal Process.*, vol. 48, no. 2, pp. 442–456, Feb. 2000. doi:10.1109/78.823971

[181] P. Strobach, "Equirotational stack parameterization in subspace estimation and tracking," *IEEE Trans. Signal Process.*, vol. 48, no. 3, pp. 712–722, Mar. 2000. doi:10.1109/78.824667

[182] J. Razavilar, F. Rashid-Farrokhi, and K. J. R. Liu, "Software radio architecture with smart antennas: a tutorial on algorithms and complexity," *IEEE J. Select. Areas Commun.*, vol. 17, no. 4, pp. 662–676, Apr. 1999. doi:10.1109/49.761043

[183] J. H. Winters, "Optimum combining in digital mobile radio with cochannel interference," *IEEE J. Select. Areas Commun.*, vol. 2, no. 4, pp. 528–539, July 1984. doi:10.1109/JSAC.1984.1146095

[184] J. H. Winters, "Optimum combining for indoor radio systems with multiple users," *IEEE Trans. Commun.*, vol. 35, no. 11, pp. 1222–1230, Nov. 1987. doi:10.1109/TCOM.1987.1096697

[185] J. H. Winters, "Signal acquisition and tracking with adaptive arrays in wireless systems," in *43rd IEEE Vehicular Technology Conference*, May 1993, pp. 85–88. doi:full_text

[186] G. Ungerboeck, "Channel coding with multilevel/phase signals," *IEEE Trans. Inform. Theory*, vol. 28, no. 1, pp. 55–67, Jan. 1982. doi:10.1109/TIT.1982.1056454

[187] İsrafil Bahçeci and T. M. Duman, "Combined turbo coding and unitary space–time modulation," *IEEE Trans. Commun.*, vol. 50, no. 8, pp. 1244–1249, Aug. 2002.

[188] İsrafil Bahçeci, T. M. Duman, and Y. Altunbasak, "Antenna selection for multiple-antenna transmission systems: performance analysis and code construction," *IEEE Trans. Inform. Theory*, vol. 49, no. 10, pp. 2669–2681, Oct. 2003.

[189] İsrafil Bahçeci, T. M. Duman, and Y. Altunbasak, "A turbo coded multiple description system for multiple antennas," in *Global Telecommunications Conference*, vol. 7, Dec. 2003, pp. 4011–4015.

[190] A. Nasipuri, S. Ye, J. You, and R. Hiromoto, "A MAC protocol for mobile ad hoc networks using directional antennas," in *Wireless Communications and Networking Conference*, vol. 3, Sept. 2000, pp. 1214–1219.

[191] K. Young-Bae, V. Shankarkumar, and N. H. Vaidya, "Medium access control protocols using directional antennas in ad hoc networks," in *Nineteenth Annual Joint Conference of the IEEE Computer and Communications Societies (INFOCOM 2000)*, vol. 1, Mar. 2000, pp. 13–21.

[192] "IEEE Standard for Wireless LAN Medium Access Control (MAC) Physical Layer (PHY) Specifications, IEEE Std. 802.11," Nov. 1997.

[193] R. Govindarajula, "Multiple access techniques for mobile ad hoc networks," Master's thesis, Arizona State University, 2001.

[194] A. J. Paulraj and E. Lindskog, "A taxonomy of space–time processing for wireless networks," *IEE Proc. Radar Sonar Navig.*, vol. 145, no. 1, pp. 25–31, Feb. 1998. doi:10.1049/ip-rsn:19981807

[195] I. E. Telatar, "Capacity of multi-antenna Gaussian channels," Bell Laboratories, Lucent Technologies, Tech. Rep., Oct. 1995. [Online]. Available: http://mars.bell-labs.com/papers/proof

[196] G. J. Foschini and M. J. Gans, "On limits of wireless communications in a fading environment when using multiple antennas," *Wirel. Pers. Commun.*, vol. 6, no. 2, pp. 311–335, Mar. 1998. doi:10.1023/A:1008889222784

[197] V. Tarokh, N. Seshadri, and A. R. Calderbank, "Space–time codes for high data rate wireless communication: performance criterion and code construction," *IEEE Trans. Inform. Theory*, vol. 44, no. 2, pp. 744–765, Mar. 1998. doi:10.1109/18.661517

[198] A. Stefanov and T. M. Duman, "Turbo-coded modulation for systems with transmit and receive antenna diversity over block fading channels: system model, decoding approaches, and practical considerations," *IEEE J. Select. Areas Commun.*, vol. 19, no. 5, pp. 958–968, May 2001. doi:10.1109/49.924879

[199] S. Talwar, "Blind space–time algorithms for wireless communications," Ph.D. dissertation, Stanford University, Scientific Computing and Computational Mathematics, Jan. 1996.

[200] A.-J. van der Veen, S. Talwar, and A. Paulraj, "A subspace approach to blind space–time signal processing for wireless communications," *IEEE Trans. Signal Process.*, vol. 45, no. 1, pp. 173–190, Jan. 1997. doi:10.1109/78.552215

[201] L. Tong, G. Xu, and T. Kailath, "Blind identification and equalization based on second-order statistics," *IEEE Trans. Inform. Theory*, vol. 40, no. 2, pp. 340–349, March 1994. doi:10.1109/18.312157

[202] E. Moulines, P. Duhamel, J. Cardoso, and S. Mayrargue, "Subspace methods for blind identification of multichanel FIR filters," *IEEE Trans. Signal Process.*, vol. 43, no. 2, pp. 516–525, July 1995. doi:10.1109/78.348133

[203] H. Liu, G. Xu, L. Tong, and T. Kailath, "Recent developments in blind channel equalization: From cyclostationarity to subspaces," *Signal Process. (Elsevier Press)*, vol. 50, pp. 83–89, 1996. [Online]. Available: http://danube.ee.washington.edu/downloadable/hliu/survey_tong2.pdf doi:10.1016/0165-1684(96)00013-8

[204] L. Tong and S. Perreau, "Multichannel blind identification: From subspace to maximum likelihood methods," *Proc. IEEE*, vol. 86, no. 10, pp. 1951–1968, Oct. 1998. doi:10.1109/5.720247

[205] S. Talwar, M. Viberg, and A. Paulraj, "Blind estimation of multiple co-channel digital signals using an antenna array," *IEEE Signal Process. Lett.*, vol. 1, no. 2, pp. 29–31, Feb. 1994. doi:10.1109/97.300310

[206] S. Talwar and A. Paulraj, "Recursive algorithms for estimating multiple co-channel digital signals received at an antenna array," in *Proc. Fifth Annual IEEE Dual-Use Technologies and Applications Conference*, May 1995.

[207] R. Steele and L. Hanzo, Eds., *Mobile Radio Communications*, 2nd ed. New York: Wiley, Oct. 1999.

[208] S. R. Saunders, *Antennas and Propagation for Wireless Communication Systems*. New York: Wiley, Sept. 1999.

[209] A. J. Paulraj, D. A. Gore, R. U. Nabar, and H. Bölcskei, "An overview of MIMO communications a key to gigabit wireless," *Proc. IEEE*, vol. 92, no. 2, pp. 198–218, Feb. 2004. doi:10.1109/JPROC.2003.821915

[210] R. B. Ertel, P. Cardieri, K. W. Sowerby, T. S. Rappaport, and J. H. Reed, "Overview of spatial channel models for antenna array communication systems," *IEEE Pers. Commun. Mag.*, vol. 5, no. 1, pp. 10–22, Feb. 1998. doi:10.1109/98.656151

[211] A. G. Burr, "Channel capacity evaluation of multi-element antenna systems using a spatial channel model," in *Proc. of AP 2000*, Davos, Switzerland, Apr. 2000.

[212] W. Schüttengruber, A. F. Molisch, and E. Bonek, "Tutorial on smart antennas for mobile communications," Vienna University of Technology, Tech. Rep., 2001. [Online]. Available: http://www.nt.tuwien.ac.at/mobile/research/smart_antennas_tutorial

[213] S. U. Qureshi, "Adaptive equalization," *Proc. IEEE*, vol. 73, pp. 1349–1387, Sept. 1985.

[214] D. Giancola, U. Girola, S. Parolari, A. Picciriello, and U. Spagnolini, "Space–time processing for co-channel interference rejection and channel estimation in GSM/DCS systems," in *Proc. International Symposium on Sygnals, Systems, and Electronics (ISSSE)*, Pisa, Italy, Sept. 1998, pp. 152–155.

[215] J. C. Liberti, Jr. and T. S. Rappaport, "Analytical results for capacity improvements in CDMA," *IEEE Trans. Veh. Technol.*, vol. 43, no. 3, pp. 680–690, Aug. 1994. doi:10.1109/25.312781

[216] M. K. Varanasi and B. Aazhang, "Optimally near-far resistant multiuser detection in differentially coherent synchronous channels," *IEEE Trans. Inform. Theory*, vol. 37, no. 4, pp. 1006–1018, July 1991. doi:10.1109/18.86994

[217] J. Ramos, M. D. Zoltowski, and H. Liu, "Low-complexity space–time processor for DS-CDMA communications," *IEEE Trans. Signal Process.*, vol. 48, no. 1, pp. 39–52, Jan. 2000. doi:10.1109/78.815477

[218] A. F. Naguib, A. Paulraj, and T. Kailath, "Capacity improvement with base-station antenna arrays in cellular CDMA," *IEEE Trans. Veh. Technol.*, vol. 43, no. 3, pp. 691–698, Aug. 1994. doi:10.1109/25.312780

[219] A. J. Viterbi, *CDMA: Principles of Spread Spectrum Communication.* Reading, MA: Addison-Wesley, Apr. 1995.

[220] B. Suard, A. Naguib, G. Xu, and T. Kailath, "Performance analysis of CDMA mobile communication systems using antenna arrays," in *Proc. ICASSP*, vol. VI, Apr. 1993, pp. 153–156.

[221] H. Liu and M. D. Zoltowski, "Blind equalization in antenna array CDMA systems," *IEEE Trans. Signal Process.*, vol. 45, no. 1, pp. 161–172, Jan. 1997. doi:10.1109/78.552214

[222] B. H. Khalaj, A. Paulraj, and T. Kailath, "2D RAKE receivers for CDMA cellular systems," in *IEEE Global Telecommunications Conference (GLOBECOM)*, vol. 1, 28 Nov.–2 Dec. 1994, pp. 400–404.

[223] D. Gesbert, L. Haumonté, H. Bölcskei, R. Krishnamoorthy, and A. Paulraj, "Technologies and performance for non-line-of-sight broadband wireless access networks," *IEEE Commun. Mag.*, vol. 40, no. 4, pp. 86–95, Apr. 2002.

[224] G. D. Golden, C. J. Foschini, R. A. Valenzuela, and P. W. Wolniansky, "Detection algorithm and initial laboratory results using V-BLAST space–time communication architecture," *IEEE Electron. Lett.*, vol. 35, no. 1, pp. 14–16, Jan. 1999. doi:10.1049/el:19990058

[225] D. Chizhik, F. Rashid-Farrokhi, J. Ling, and A. Lozano, "Effect of antenna separation on the capacity of BLAST in correlated channels," *IEEE Commun. Lett.*, vol. 4, no. 11, pp. 337–339, Nov. 2000. doi:10.1109/4234.892194

[226] V. P. W. Wolniansky, G. J. Foschini, G. D. Golden and R. A. Valenzuela, "V-Blast: An architecture for realizing very high data rates over the rich-scattering wireless channel," *Proc. URSI International Symposium on Signals, Systems, and Electronics (ISSSE '98)*, Pisa, Italy, Sept. 1998, pp. 295–300. doi:full_text

[227] P. Mannion, *Communication Systems Design*. Paul Miller, June 2002, ch. Smart Basestations Maximize Capacity, pp. 15–20.

[228] "Metawave Communications," 2002. [Online]. Available: http://www.metawave.com

[229] "ASCOM AR&T." [Online]. Available: http://www.ascom.ch

[230] J. Lu and T. Ohira, "Smart antennas at wireless mobile computer terminals and mobile stations," in *IEEE AP-S International Symposium and USNC/URSI National Radio Science Meeting*, Boston, MA: July 8–13 2001.

[231] G. T. Okamoto, *Smart Antenna Systems and Wireless LANs*. Dordrecht, Netherlands: Kluwer, 1999.

[232] R. H. Roy, "An overview of smart antenna technology: the next wave in wireless communications," in *IEEE Proc. Aerospace Conference*, vol. 3, Mar. 1998, pp. 339–345.

Printed in the United States
by Baker & Taylor Publisher Services